Multivision

Multivision

The Planning, Preparation and Projection of Audio-visual Presentations

by
John Lewell

Focal Press · London

Focal/Hastings House · New York

© FOCAL PRESS LIMITED 1980

  British Library Cataloguing in Publication Data

Lewell, John
    Multivision.
    1. Magnetic recorders and recording
    2. Slides (Photography)
    I. Title
    621.389'32   TK7881.6

ISBN (excl. USA) 0 240 51026 7
ISBN (USA only) 0 8038 4728 9

Printed and bound in Great Britain by
M & A Thomson Litho Ltd., East Kilbride, Scotland.

Contents

power projection; Fade/dissolve xenon projectors;
Projection lenses; Basic projection optics; Condenser
systems; Projector lenses; Calculating projection distances;
Calculating correct size of lens; Calculating picture width
and height; Slide apertures; Types of lenses; Projector
mounting; Image illuminance; Subjective tests.

Presenting the script; The storyboard; The soundtrack;
Visual effect; Work on the storyboard; Plotting tension.

showing; A good start; Starting out; Scripting; Photography; Graphics; Handling the slides; Putting the show together; Encoding; The sound laboratory; The communications room; Designs for audiovisual; The control panel.

Acknowledgements

The author would like to thank all friends and colleagues in the audio visual industry who have contributed either information or encouragement, or both, while this book was being written.

Colour plates
Sixth Bahman Museum, Madame Tussaud's, Philips Evoluon: *Robert Simpson and Denis Naisbitt, Electrosonic Ltd.*
St Michael's Church: *Robin Prater, Prater Audio Visual, Greenwich.*
Washington National Visitors' Center: *Gregor Grieg, Media Generalists, San Francisco.*

Acknowledgments

Introduction

A multivision show is not a film. It may have many of the elements that make up a film such as a mixed soundtrack, a succession of images, a narrative, and even a development of characters within that narrative – but it is still not a film. Conceiving and producing a multivision show requires a new discipline, a considerable mastery of the new technology at the producer's disposal, and the ability to use this technology in a way which is unique to the medium.

It is true that any new medium, indeed any new invention, bears a striking resemblance to its immediate predecessors, until, with use and development it has found its own identity. Early motor cars looked like horse drawn vehicles minus the horse – and early television could have been mistaken for attempts at filming radio programmes. Multivision has undergone a similar development.

The production of slide transparencies using the most sophisticated cameras, emulsions and processing had been brought to a pitch of ingenuity and quality before anyone gave much thought to actually looking at them. If you projected them one at a time on to a white screen you would see them quite adequately and that might have been the end of the story. After all, film; which is a long strip of transparencies' projected intermittently; gives a two-dimensional representation of real life, and the almost perfect illusion of movement. Why was there any need for a medium in between these two extremes?

The answer is that these media were just that – extremes. The slide show of unrelated pictures accompanied by the live commentary of the photographer became synonomous with boredom, and movie making became the epitome of vast expenditure, with at the very least a *cost* of thousands. In between lay a vast market of people who wanted to communicate, promote their products, educate their staff, even to entertain the public using visual images accompanied by prerecorded sound. Hence the birth of audiovisual presentation.

1

Following its birth audiovisual rapidly grew into a monster. It is now possible within a given environment to call up any image, whether moving or static, accompanied by any sound, at any time, and in any part of that environment. It has, in a word, lacked discipline because of the very freedom which it allows.

The purpose of this book therefore is not only to enumerate and explain the techniques which are available to the producer; I have also attempted to define the identity of the multivision medium and to demonstrate that the most successful multivision shows have a unique combination of ingredients. If this book encourages the viewer to be more critical, the sponsor to be more demanding and the producer to be more disciplined then it will have succeeded in its task.

1 Multivision, slide tape, the multiple medium

Each of the visual media has evolved through a variety of formats before one or two shapes and sizes have been chosen as standard. The word 'format' normally refers to the picture shape that the viewer actually sees. However, 16mm 'academy' movie and 35mm 'academy' movie give the same format on the screen and yet they cannot be played on the same projector. They are different *standards*, and inherent in the word 'standard' is the concept of quality – which is not inherent to the word format.

Formats

With video and television, for instance, formats and standards are more defined. The format is always the shape of the universal television screen, although in viewer terms this may be a Sinclair 2in pocket TV or a large Eidophor projected picture. The TV format is unique in being universal and broadcast television has greatly influenced the new video industry.

Even so, videotape machinery is nowhere near standardized. Standards include $\frac{1}{4}$in Akai, $\frac{1}{2}$in Philips, $\frac{3}{4}$in Sony, 1in IVC and 2in broadcast standard. Each of these tape widths is associated with a particular combination of electronics to record and reproduce a programme. It is as though all magazines and newspapers were of A4 shape and yet required special viewing equipment costing hundreds of pounds to decode the words printed on them.

Film and video standards and formats are essential because the whole business depends on exchange; and on distribution to a mass audience in exchange for money. Film has been a mass market medium from its beginnings and video manufacturers have had one eye fixed on the same market.

How much less standardization can we expect from multivision? It can be argued that multivision remains a relatively small business because it does not have a mass market. It could equally be argued that the differing formats and standards are so numerous that the means of distribution of programmes can never be achieved.

So how many formats and standards have been used in multivision? They are too numerous to list. Not only does the visual format (equivalent to the TV screen) have dozens of different shapes, but there is the same diversity in the electronics that are used for replay and in the projected material itself.

Leaving aside the question of electronics, which is explained later, let us examine the almost abstract question of the format of what the viewer sees.

Picture shape

At its simplest multivision may be a single-screen show with two projectors and mono sound. The screen ratio can be the 35mm film format; that is to say in the ratio of 2 high to 3 wide (from the 24 × 36mm dimensions of an ordinary slide). Alternatively, the slide may be masked to give any shape or size of format; although I have not yet seen a single screen show where *all* the slides are triangular or circular. A useful alternative for some subjects is to use the 35mm slide in a vertical position: 36mm high and 24mm wide. That is *portrait* instead of *landscape*. A programme entirely about tower cranes or high rise buildings would be more easily photographed in this format for projection.

In still photography, a square format is popular because some of the highest quality cameras take square pictures. The existing state of photography is very much the father of multivision. In some work, photographing fashion shows, for instance, 60×60 mm ($2\frac{1}{4}$ in square) cameras are almost universal, so it is sensible to retain the square format for projection.

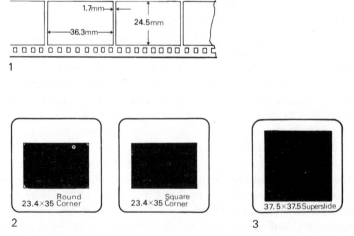

Multivision slide mount standards. 1, Frame dimensions for film exposed in a conventional 35mm camera. This is the standard unmasked size for 35mm slides. 2, Standard apertures in Weiss Registration Mounts for 35mm film. 3, For sprocketed 46mm film or use film cutting unit on sheet film.

4

Since there is difficulty in obtaining 60 × 60 mm projectors, square format pictures may be reduced to superslide format. That is, 38 × 38 mm which retains the 50 mm (2 in) mount. This slide mount is the one universal standard that has emerged in the multivision business.

Superslide has the advantage of using the maximum area of a transparency within the 50 mm mount. This makes for a relatively higher definition on the screen. It should be remembered, though, that not all projector lenses can cope with the increased area of superslide.

Square format is also neutral in its emphasis on verticals and horizontals and many producers choose it for this reason. Having said that, however, it is interesting to note that film producers have tackled nearly every subject imaginable within an unvaryingly horizontal format. The only exception is the early Russian cinema with its characteristic high picture frame that, in any case, is lost when projected on to Academy screens in the West.

Irrespective of the slide shape, a square screen is often chosen for single-screen presentations. It can accommodate a range of special shapes, or be masked down to provide horizontal or vertical 35 mm slides.

Cinema screening

It may be necessary to play your single-screen show in a fixed cinema installation. Although most cinemas are equipped with movable motorized masking, the wide-screen format is widely used and the ratio may be 2:1 or $1\frac{3}{4}$:1. In this case, you can make or buy special slide masks. For example, you may choose the Academy format itself. This ratio is 4:3 or somewhere halfway between superslide and landscape.

Further variations

Already for the simple single-screen show we have listed six different shapes of projected image. With a choice of mono or stereo sound that is twelve different possibilities. What is more, this does not take into account the ingenuity and inventiveness of those producers who can find even more combinations. We have chosen to ignore single projector shows (univision, monovision?) because they do not give a continuous picture. So we come to the combinations that two projectors allow.

The two can be offset. That is, each projector may be pointed at a different area with, say, 50% of the picture being common to both. This novelty is frequently used in exhibitions; although it is more often seen in shows that use more than two projectors.

Photographically, the image may be split up to give the effect of a large multi projector show. By using lots of superimpositions – both projectors on at the same time – this can create various interesting effects. It can be overdone though sometimes. The time and trouble that is spent creating these effects could be better used on other aspects of the show, perhaps on the soundtrack or the message or the script itself. There are no short cuts to good multivision; and there is little point in trying to emulate the effects of a bank of projectors if you are working with only two.

Three projectors

Adding a third projector to the show is often desirable from the producer's point of view. It enables captions to be kept together in one magazine and it increases the overall slide capacity. However, few presentation systems and two projector dissolve units have the option of adding a third projector. In practical terms, you have to jump into a more complex class of electronics and, proportionally pay considerably more than one-third extra.

More projectors

Having four, five or six projectors covering the same screen area brings a number of benefits to the show, and a few drawbacks. It increases slide capacity greatly, and makes a more rapid rate of picture change possible. You can create animated effects and more complex visual changes.

Some of the most effective uses of multivision have been single-screen multiprojector (i.e. four or more projector) shows. This format (dare I use the word) allows a great variety of slide masking, setting pictures within other pictures, changing both sides of the image area independently while bringing on captions; all of these tricks without having to force the viewer's head to swivel toward another screen area. The whole presentation is achieved within an area that is comfortable to the eye.

The single-screen multiprojector show may not be as spectacular as a giant multiscreen but it can impart just as much visual information. And unless there is communication involved, multivision would be of little use to anybody. However, the weight and bulk of projection equipment are important considerations. If you have chosen to limit your multivision show to a single screen it is likely that the reason is so that it can be easily transported, or perhaps it is being shown in a confined space. Maybe, also, your budget is limited, especially if you want to duplicate the show and have it presented simultaneously in a number of places. There is also the increased difficulty of aligning six projectors, each with its own lens, so as to give the ideal visual impression that the image is coming from a single source. This is by no means insuperable, given a sufficient projector throw; but it is the reason that six projectors is usually the maximum used on a single-screen area.

Two screens

In terms of screen areas, the first step up from single screen is two screen. This

6

can be an uncomfortable format to watch because equal emphasis is given to each screen. The viewer will look from screen to screen as each changes its picture; giving rise to the 'Wimbledon effect' which is often an observed phenomenon in multivision audiences. If both screens change at the same time, each showing a different image, then the eye is confused; or the viewer has to make a conscious decision about which screen to look at.

Two screen shows are born out of particular situations. Firstly, the user may not think that he has enough money for a classic three-screen installation. Secondly, his theatre or viewing area may not easily accommodate a three-screen side-by-side matrix. Thirdly, he may want to show visual comparisons on the screen, (before and after pictures; this year's figures and last year's figures). This chapter deals mainly with the visual abstractions of multivision rather than with its applications. So we must not overlook the fact that a two-screen show can be highly attractive. For instance, a two-screen show provides a particularly splendid single image of, say, a landscape. The overall ratio is then 3:1 which is pleasing to the eye. What must be pointed out though, is that the natural starting point for the eye, even before the show begins is the centre of the screen. With the two-screen show this is the dividing line between the two areas. So I should make one rule at this point: *never* use two screen with a wide division between the screen areas. After all, the aesthetics of multivision must be attractive before it can communicate anything.

Again, you can have two-screen areas with two projectors on each; or up to six projectors on each. You *can* overlap the central dividing line with some of these projectors. As with single screen, the extra flexibility increases the complexity of producing the photography and increases the difficulty of line up before the presentation of the show.

Three screens

Moving on then to use of three screens, we come to what I shall call a classic matrix. 'Classic' is a word which can be applied to a medium or style when traditionally it has been seen to have a 'rightness' about it, lacking in other styles by comparison.

To begin with, emphasis is naturally given to the centre screen, and this may be reinforced by the addition of extra projectors. The images on either side can support the central image and enhance the overall effect. In mixed media shows it is common to reinforce the centre screen with movie projection, with the option of using slides on the side screens at the same time.

The only real disadvantage of the three-screen concept is that most theatres designed for cinema projection are too long and narrow to accommodate the extreme ratio 4·5:1. The people at the back of the audience will not have a particularly good view, since it is the height of each individual screen in relation to the distance of the viewer that is the critical ratio in theatre design.

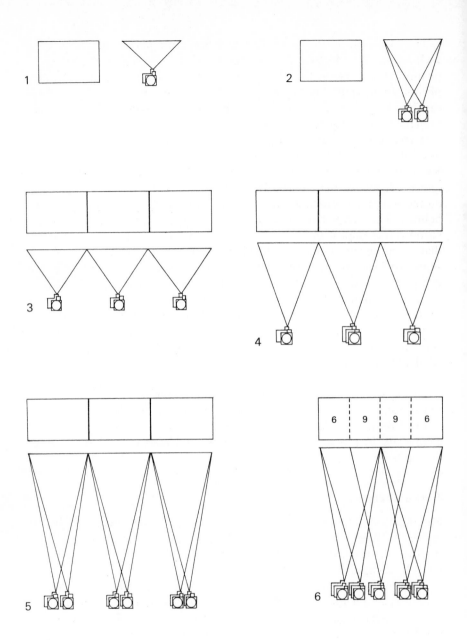

Screen formats—a selection. 1, Single-screen. Two projectors. 2, Single-screen. Four projectors. 3, Three-screen. Two projectors per screen. 4, Three-screen. 2-3-2 projectors. 5, Three-screen. 4-4-4 projectors. 6, Two-screen with overlapping third screen. 7, Three-screen with soft-edge mask overlap. 8, Six-screen. Three projectors per screen. 9, Seven-screen. Cine on centre three screens. 10, Unusual formats. Screens of varying sizes.

8

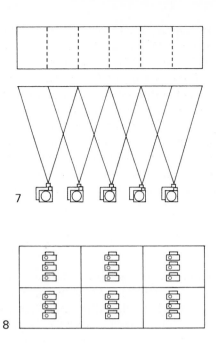

7

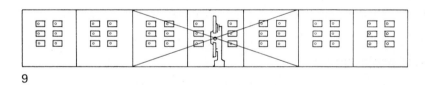

8

9

10

At this point in discussing screen formats I must look ahead briefly to topics discussed later in greater depth. If you are using dissolve units that control two projectors in unison, that is, upon receiving a single command signal the dissolve unit dims the lamp on one projector and fades up the lamp on the other, then it is inadvisable to attempt overlapping projector formats. For this, you need individual projector control. With full control of each projector it is quite feasible, in the larger system, that is three screen and upwards, to achieve otherwise impossible visual effects by overlapping images. Chief among these is the ability to make an image appear to move from one side of the (whole) screen to the other – and back, if required. This visual effect has a unique way of integrating the very wide 4·5:1 ratio of three screens and 7·5:1 ratio of five screens.

More and more

In the USA, the technique of overlapping is so widely used that three-screen, five-screen and seven-screen (all in-line) have become the most commonly used. In Europe multivision has developed along different lines, using dissolve units, and giving rise to three-screen, six-screen (3 wide × 2 high), nine- (3 × 3), twelve- (4 × 3), and twenty-four-screen (6 × 4). In the European tradition it is easier to see how the effects are achieved and there is less dependence on the magic or gimmickry of the medium – depending on your attitude to overlapping formats. Now that individual projector control is universal and one might say 'de rigeur' for the serious producer, it will be interesting to see the differing approaches in various parts of the world.

Rear projection

If rear projection is used then it will continue to make sense to have each group of projectors covering the same screen area, with adjacent screens separated by batters ('eggboxing' – see glossary) in order to avoid an ugly overspill of light from one slide to another. The most accurate lining up and registration will not prevent a slight overlap, especially if several projectors are used. This can only be cured by a physical dividing strip against the screen itself.

Varied image sizes

In discussing overlapping formats I have assumed that each picture image projected is of exactly the same size. That is, that all the slide mounts are similar, lenses of the same focal length are being used, and all projectors are equidistant from the screens. Now, many producers have used a mixture of lenses to give images of different sizes. For instance, a show made for the

British Royal Navy a few years ago showed a changing series of sea-scapes upon which were superimposed images of ships, carriers and missiles that appeared to be moving. The sea-scapes were projected with a wide-angle lens over the whole screen area using two projectors, whereas the other images, which included emblems and flags of the fleet, were shown by overlapping projectors. The technique worked reasonably well, because superimpositions need to be burnt in with a more intense light onto a darker picture. Occasional rapid movement was produced by a fast ripple sequence of overlapping slides showing, for instance a missile being launched. However, the visual content of the show was limited by the format. The projectors covering the whole screen could not show anything *except* a sea-scape and this had to be retained for the whole show. Well, I guess that's what you join the Navy for. . .

The major problem of using lenses of different focal lengths in order to obtain different screen formats of varying sizes is simply that the light when spread over a wider area appears to be diminished. I much prefer to see each projector balanced with all the others. The alternative solution is to mask down the slides themselves in order to create screen areas of different sizes. Some producers have suggested that this is not necessary because their photography is specially taken in order to balance the effect on the screen. In practice, this is almost impossible and only trial and error will ensure a good result. Yet another variable is in the lenses themselves. Longer focal length lenses compensate to some extent because they transmit less light.

In a properly constructed multiscreen show it should be possible to place identical slides of uniform density in the gate of each projector and to retain a light meter reading that does not vary greatly on any part of the screen. This is a test which is rarely carried out.

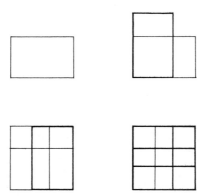

Using portrait and landscape formats.

Unusual shapes

Moving now away from classic formats into the realm of the unusual – and even the downright eccentric – we come to combinations of slide maskings. For instance, interlocking hexagonals can make an attractive centre-piece for an exhibition stand, although here one must be prepared for the format to be obtrusive. Throughout this type of show every change of slides carries the message, 'you are looking at seven very cleverly interlocking hexagonal slides'. If this is acceptable, the possibilities are limitless. Triangles, circles, semi-circles, dodecahedrons. . . why not?

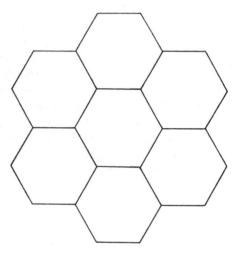

Hexagonal screens.

At a very early stage in the development of multivision, producers realized that the screen itself did not have to be in the same flat plane. A wide seven-screen show is easier to view if the screens are gently curved. Taken a stage further, the screens can complete a circle, right round the auditorium. This does mean that the audience is not able to see the majority of the show but at least the sponsor and the producer can have a lot of fun making it. The audience will see a spectacle, like being at a fun-fair. If this is the intention, it is another good use of multivision.

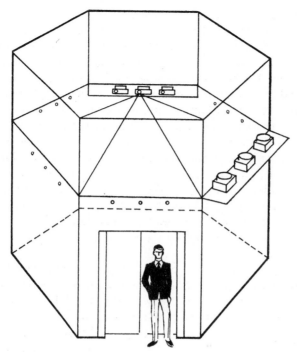

Projection in-the-round.

Mobile screens

One of the most spectacular installations has been the sixth Bahman Museum in Tehran. This carries the concept of experiment a step further, for not only are the screens not in the same plane, they move about during the show! The format is a modular one. Each screen is square and has two projectors. However, they are stacked one above the other to form a huge column which is suspended from a track in the ceiling. Each column is enclosed in order to carry up to 5 dissolve units and 10 projectors behind the screens. There are 12 such columns and during the show they can move to either side while a 35 mm film automatically starts on a cue from the master tape.

Adding movies

Integrating film with slides requires a considerable amount of care. In the Tehran show this is solved by separating the two media, projecting them on different screens entirely. Many people, even regular cinema goers may view a multivision without appreciating just how much better the picture quality is.

13

Take, for instance, a twelve screen show, made up of 35 mm slides. The actual emulsion area projected is 10 368 sq. mm. The area of a cinema film that is projected intermittently, once every twenty-fourth of a second is 432 sq. mm. It is likely that the emulsions used are very similar. Of course, the comparison is slightly unfair. The brains link together the images of a film (by persistence of vision) and it concentrates on the action, so the actual material quality of the image is less important. In multivision the viewer often has time to criticise and experience the quality of the pictures, besides seeing their content and aesthetic merit. So, the images must be extremely high quality.

Photographically, a good quality slide will have a far greater range of tones than the average 35 mm movie print. It would be a mistake to show the two media side by side without compensating to even out the differences. Therefore it makes sense to tone down the slides that are to be shown while the film is running. It is no use using beautiful chocolate box shots, mastered on 10 × 8 in. film, on the two outside screens while showing a 16 mm film on the centre screen.

Even with quite exceptional print quality, the audience will think that you have not made a very good film. The slides are *too* good by comparison. No audience is likely to grumble afterwards that 'the slides were far too good'. I am not suggesting that faults should be built-in to the slides or that they should be out-of-focus or overcontrasty. Rather, they should show material with short tonal scales. In any case, the audience will watch the moving image for this may reveal a surprise in a fraction of a second which would otherwise be lost.

Sound

Before attempting to recap the variety of formats listed above, a word about sound. The audio part of audiovisual is as important – arguably, more important – than the images themselves. If you are constructing a grandiose multivision in a giant auditorium then you have to match the images with equally grandiose sound effects.

A three-screen show needs stereo sound unless it is being shown in an area where quality audio is impossible. Just as one should ensure that the light on adjacent screens is balanced so you should ensure that your sound balances with the visual effect. Wherever the budget allows, a large multivision should be accompanied by 25 mm (1 in.) sound replay equipment. This gives a total of eight independent tracks: one for control, one for a clock track, one for lighting effects and, say, five for sound. Loudspeakers can then be placed close to the different screens and it is then possible to locate a particular sound physically with a particular image. While this can sometimes be left to the imagination of the audience, I believe that a good production will, from the very start, be conceived as a succession of sounds and visual images that as a whole are greater than the sum of their parts.

14

Lighting

The control of lighting is frequently desirable during a multiscreen presentation. In any case, it is essential to dim houselights at the beginning of any projected show, even on an exhibition stand or in a boardroom. This should be a part of the show itself and most well-engineered systems include this facility as a function of the playback equipment.

However, there are many occasions when the show itself extends beyond the screens and into the auditorium itself. In a sales conference, for instance, if the product is reasonably small and can be physically brought into the theatre it should be. It would be odd to merely show photographs of it to the sales force. The audio visual programme can show, better than any other media, close-ups of the controls, a description of manufacturing, facts and figures related to marketing, diagrams of circuitry and so on. But, after all, it is the actual object that the dealers have got to stand on their shelves. So as part of the audiovisual show the product can be spotlighted, on cue.

An architect's presentation can benefit greatly from this technique. Three-dimensional models can be lit from inside or by spotlight to show that part of the building scheme which is currently being discussed on the multivision itself.

. . . and more

The master tape (which controls the slide changes) can cue not only 'spotlights but other effects, such as smoke generation. For example, it is quite feasible to cue an electrically operated atomizer to spray a product into the theatre. That is ideal for a perfume manufacturer's new product launch. Another cue would start a ventilation system to clear the air for the next brand.

These techniques take multivision into the realms of theatre. In conference shows, stage personalities may take part in person and their role has to be integrated into the audiovisual display. This is as difficult as mixing movie film with slides. The brain applies different rules to each of the different media.

Personally, in mixed-media shows I like to see an attempt at creating a new style, almost a new medium. One that is not the typical variety show with a link man and a dozen different acts. Because there, with each new act, the audience has to adjust its expectations. If you see a trapeze artist, you expect to be amazed by skill, grace and daring. You do not expect him to make you laugh. In the same way, an audience will have expectations when they see a live presenter of a conference and different ones when they watch an audiovisual show. Suppose, though, the presenter gives an 'off-the-cuff' live commentary to the show (it may well have been rehearsed at great length until the throw-away remarks are perfectly timed). Then I believe one has the beginning of a new experience. It is possible to inject drama into a slide show and create a piece of theatre.

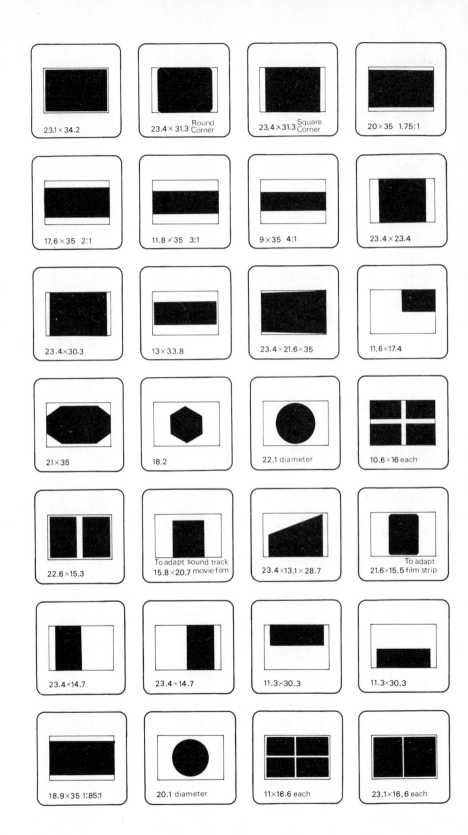

23.1 × 34.2

23.4 × 31.3 Round Corner

23.4 × 31.3 Square Corner

20 × 35 1.75:1

17.6 × 35 2:1

11.8 × 35 3:1

9 × 35 4:1

23.4 × 23.4

23.4 × 30.3

13 × 33.8

23.4 × 21.6 × 35

11.6 × 17.4

21 × 35

18.2

22.1 diameter

10.6 × 16 each

22.6 × 15.3

To adapt sound track
15.8 × 20.7 movie film

23.4 × 13.1 × 28.7

To adapt
21.6 × 15.5 film strip

23.4 × 14.7

23.4 × 14.7

11.3 × 30.3

11.3 × 30.3

18.9 × 35 1:85:1

20.1 diameter

11 × 16.6 each

23.1 × 16.6 each

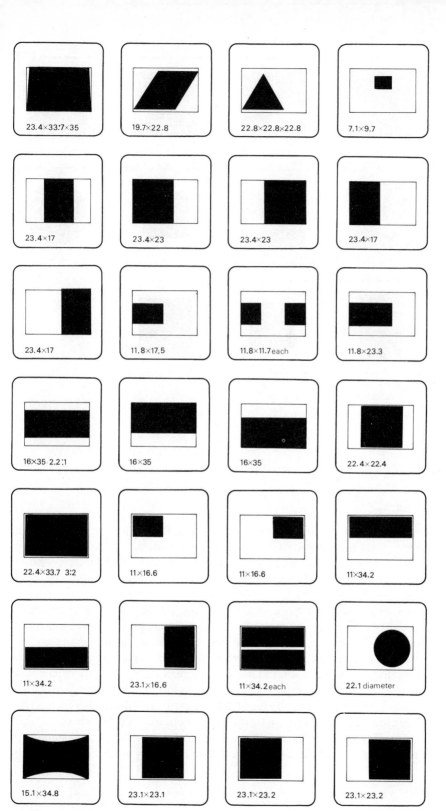

23.4×33:7×35	19.7×22.8	22.8×22.8×22.8	7.1×9.7
23.4×17	23.4×23	23.4×23	23.4×17
23.4×17	11.8×17.5	11.8×11.7 each	11.8×23.3
16×35 2.2:1	16×35	16×35	22.4×22.4
22.4×33.7 3:2	11×16.6	11×16.6	11×34.2
11×34.2	23.1×16.6	11×34.2 each	22.1 diameter
15.1×34.8	23.1×23.1	23.1×23.2	23.1×23.2

Special aperture slides.

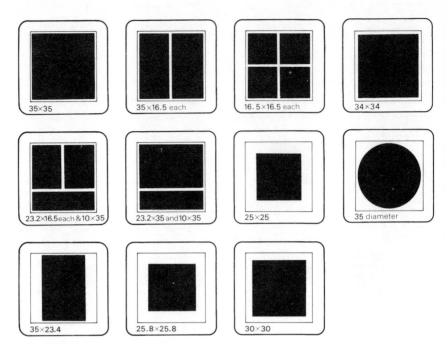

The figures are labelled: 35×35; 35×16.5 each; 16.5×16.5 each; 34×34; 23.2×16.5 each & 10×35; 23.2×35 and 10×35; 25×25; 35 diameter; 35×23.4; 25.8×25.8; 30×30.

Special aperture superslides.

Uses

Inevitably, a discussion of formats leads to all the many uses of multivision, to the extent that format itself almost disappears during the invention of new media experiences. But without a degree of discipline, without a recognition of the aesthetic boundaries of a medium, it is impossible to attain the excellence which all producers, artists, writers—try to achieve. Although in mixed-media shows I think experiment with the basic concepts are vital, I should prefer to see the majority of multivision shows keeping to, say, a dozen basic formats. The creative task of reinventing the language every time you want to speak is too onerous to be even remotely possible. I have no doubt, for instance, that many single-screen shows are better than most multiscreen shows. Simply because the techniques have been mastered, allowing so much more to be suggested and evoked by the images themselves.

Multivision, then, has one universal standard: the 50 mm (2 in.) slide mount. Beyond that it has common features, but very few real standards. Programmes for single-screen presentation units that are made by the major

18

manufacturers are interchangeable. It is commercially viable to make such a programme and issue it on a particular format and have a reasonably good chance of making a profit. However, large multivision ventures are one-off spectaculars made for a specific purpose to reach a specific audience. They have to justify their cost on this one-off basis. For this reason there is no multivision equivalent to Hollywood. Production companies are, frequently, small firms consisting of one or two enthusiasts and their assistants who are willing to tackle each job with a flair for invention and experiment.

2 Applications of multi-vision

When a new communications medium emerges, not all of its potential applications can be foreseen. It has to grow very slowly. New 'languages' have to be developed. The past history of communications has to be searched for old ideas that have been lost and need to be rediscovered. Indeed, it takes great creative minds to demonstrate that the medium has the power of expression which is associated with other, competing, media.

The audiovisual medium has not yet achieved the potential which, to judge by its component parts, it should be able to attain. Let us look for a moment at those component parts. Each one is a medium in itself. Each is capable of carrying the most mundane—or the most elevated—expression or meaning.

Components of audiovisual presentations

The still photographic image This can be either realistic or abstract; specific or mysterious; a representation, strictly an 'interpretation', of another art form (such as painting, drawing, sculpture, architecture); or symbolic, a word or series of words, arithmetical figures or signs.

Human speech This can be spoken prose; spoken poetry; recorded conversation; drama; or even oratory.

Music The least utilitarian medium but certainly the most evocative of emotion. The most 'formal' medium, appealing to our sense of harmony and order.

Other audible sounds These can be either abstract sounds or sounds resulting from some activity which will be imagined by the listener on hearing them.

There is, therefore, an embarrassment of riches. Small wonder that no one has yet used them to their full effect by creating an audiovisual show which is acclaimed as a masterpiece. Now, you may object and say that the film medium has all of these ingredients, plus movement, and that many films can rank as some of the most distinguished expressions that man has ever created. This is quite true. Film has the tremendous advantage of being able to record human beings walking and talking, interacting with each other and displaying a wide range of human activity. Film can also record non-human processes and activities. Multivision uses a quite different approach: the presentation of high quality individual images. These images switch, merge, superimpose and augment one another to provide an experience unique to the medium. Of course, you can imagine using the whole range of multivision techniques in a film, but it will still lack the major strength of multivision: its superb image and sound quality.

I make no apology for this essentially 'highbrow' introduction to the applications. After all, how do you apply something which does not yet have a thoroughly defined function. By contrast, you can apply a knife to peeling potatoes because it has a sharp edge. You can tighten up a screw with it, or prise off a bottle top or carve a piece of wood with it. In fact, I can immediately think of more applications for using a knife than I can for the whole medium of audiovisual.

This can be explained by the fact that any tool is made for a purpose. Although it may attain a purely symbolic status, it is really a utilitarian object that is used in some human activity. It enables that activity to take place.

At present, the audiovisual presentation is assumed to be a functional tool. It is a tool for selling; for student and staff training; for persuading that A is better than X. Audiovisual has emerged as a consequence of these activities. It is a medium that helps these activities take place.

However, to return for a moment to the many components: images, speech, music and sound effects. Some of these are of great complexity and subtlety. In other words the audiovisual medium is potentially very sophisticated; but is being used in a humdrum fashion. The present uses are equivalent to using a computer for counting sheep.

In no way do I wish to denegrate the activities of selling or educating and informing people. What I do object to most strongly is the assumption that audiovisual is a *ready-made* tool for helping with these activities. To use the medium effectively, you have to begin to think audiovisually. Thinking verbally is not easy, thinking visually is more demanding and thinking audiovisually even more so.

Fortunately, however, any thinking process improves the more you do it. Many people have learnt how to program, that is encode, an audiovisual show. But we now need to learn how to program ourselves in order to conceive the show.

Let us look then at each current application of audiovisual and consider the kind of show which is most appropriate in each case. We should also consider the method of staging the show for this will vary with each application.

Business presentation

The business presentation comes top of the list because audiovisual methods are particularly suited to this purpose. Indeed, I believe business presentations are becoming more frequent and more elaborate because we now have the technology to make them more effective. But before considering the audiovisual content let us examine the presentation as a whole.

What is a business presentation and why do we have them? Most frequently they are staged by a sales or marketing department for the purposes of selling a product (or group of products) or a service. The message to be communicated in a short space of time may be quite elaborate, involving not only an explanation of a service but also information about price structures, geographical territories, growth rates and so on. An attempt to communicate such a 'package' of information requires help. Printed leaflets and brochures can be distributed, but these can be read only after the presentation. Because the printed page communicates to one person at a time, the sure way of fragmenting your presentation is to hand out information in printed form. People start reading it, and stop listening to you. This is also true of board meetings, committee meetings, any gathering for the purposes of discussion and decision making.

A presentation implies that there is one (or more) presenters, and a group of people who are listening and watching. The idea that people are watching is still relatively new. We talk of going to 'hear' a lecture, rather than going to 'see' a lecture. But if it were just a question of listening then it is now harder to justify having people there in person. There are great demands on a person's time, and an occasion, such as a presentation, promises to be more than merely an exercise in listening.

A business presentation is 'a formal introduction'. The observance of formality is very important because it means that the presenter has prepared what he is going to say and show, and in return expects the audience to watch and listen until such time as the formality is dropped. At the end of the presentation, for instance, there may be a question and answer session which is, essentially, informal. It is possible to give a formal presentation to any number of people. In practice, the formality tends to increase in direct proportion to the size of the audience. The secret with very small groups is to try and put some distance between you, the presenter, and the people to whom you are presenting, otherwise you will find yourself relating individually to each person and effectively they will be in charge.

Reasons for presentation

From the above, then, presentations are held because:
1. They make most effective use of time. They enable one person to address a number of people about a subject on which he or she is most qualified to speak.

2. They concentrate on essentials. A presentation is not the time for going into minor details which can be explained in printed matter.

3. They give a sense of occasion. There is a feeling of expectancy before a presentation. It is, after all, an 'introduction'. Something new will be seen and heard. Do not, therefore, disappoint the audience.

4. They provide an opportunity for group communication. The people who attend a presentation will have been invited because of some specialized interest which they share in common.

Having defined what a presentation is, we must also define its objectives. The sole objective is communication. If this is analyzed then we can say that it is the communication of facts, knowledge and emotion for purposes of training, informing, motivating, educating and selling. Particular presentations may place the emphasis firmly on just one of these objectives. In fact, the more defined the objective is, then the more successful the presentation is likely to be.

Control your show

If you are preparing a presentation, the one single rule is *control*. You need to have control over the *environment* in which it is given. You need control over the *structure* of your presentation and too you need to *rehearse* the whole exercise in order to ensure that you can retain control over it. Go through each of these points in turn before you finally stand in front of the audience. You will find that your confidence is increased because the preparation has been both personal and thorough.

Let us now take just one of the above key points, structure, because it is here that the audiovisual show becomes a necessity. If you can alternate your periods of speaking with short, prerecorded audiovisual shows then almost certainly you will improve the structure of your presentation. This is because you will be varying the pace and introducing a completely different dimension to the proceedings. When the show has finished, you will have the renewed and regenerated attention of the audience.

Few people can concentrate on an evenly paced delivery for long periods You cannot read a textbook from beginning to end in one sitting. But you can read a novel in one sitting because the author has control over your attention by means of alternatively relaxing and concentrating the activities which he is describing.

This variation in pace can be achieved without any time wasting. You can make some points seriously and others humourously. Each will benefit from the other.

An audiovisual show which is to be used as part of a presentation should be designed and produced by experts. If they have done their job well, the audience will tend to congratulate the presenter as though it were his own work. If you think this gives you an unfair advantage, you are quite right, it does; but remember that the opposite is also true. Present them with a shoddy or boring programme and the slow handclap is meant for you and not the producer.

In briefing the producer of a show you should bear in mind that there are certain ground rules which should be followed. A good producer will advise correctly, but do not give him a brief which forces him to sacrifice his integrity. The ground rules are:

1. Keep it short. 10–12 minutes is an average length for this type of show.

2. One show/one message. A number of facts may be communicated in a single show, but there should be only one message that binds them together.

3. Use the visual element. The audience will have already spent a long time looking at the presenter. They need visual stimulation.

4. Keep it moving. Many audiovisual shows suffer from a surfeit of commentary and a dearth of images. The average time for each image to remain on the screen should not exceed 6 seconds. This is only a guide, but I would emphasize the point strongly. A hundred slides in a 10 minute two-projector show is a useful number for which to aim.

If every one followed these simple ground rules, many of the current uses of audiovisual in presentations would be improved. The audiovisual contribution, in turn, improves the control over the ideas which you have to communicate, it improves the structure of the presentation, the process of communication itself, and it gives the presenter invaluable support.

Training seminar

As I have indicated while discussing presentations there is a strong informational element in a formal presentation. However, this is normally subordinate to the other objectives, particularly those of motivation and persuasion. The training seminar, by contrast, is almost wholly informational. The people attending such a session are already motivated by a desire to learn, although of course this interest must be sustained by imparting information in a stimulating way.

The first thing to be said about training seminars is that there are a lot of them. Every business has to train its staff. Schools and colleges are exclusively involved in training in the broadest possible sense. It is not surprising therefore that any sophisticated equipment which is used must be economical to run. Training programmes, in general, have to be produced quickly and cheaply. And it is for this reason that the use of economically produced 35 mm slides accompanied by a simple soundtrack proves to be a viable choice when compared to other media.

The uses of audiovisual in training can be divided into two categories: *individual* amd *group* communication. These categories relate specifically to the kind of equipment which is used. For individual communication a small self-contained unit with a projector and rear-projection screen is used to replay a programme. In this instance, learning is a solitary, individual activity, like reading a book.

Once you have a number of students, then the training seminar takes over and the need is for a more substantial means of showing an audiovisual programme. For reasons of continuity, you should use two projectors, with a dissolve unit and tape control.

Integrated systems

Frequently, the ideal system for training purposes is a complete record/ playback unit which contains the dissolve unit, decoder, and a cassette deck on which both the synchronizing tone and the commentary can be independently recorded. This enables a lecturer to make his own programmes which he can then use over and over again. Not only that, he can also replay professionally made shows which have been studio recorded with professionally taken photographs. Several companies make shows on topics which are likely to be relevant to many different groups of students.

Why use audiovisual in training?

The argument is often heard that teaching is what a teacher is paid to do and he should be capable of doing it without complicated electronic control of projected images. Some teachers do indeed feel that their authority is threatened if any teaching aid is introduced into the classroom. I should refute this by the following observations:

1. An audiovisual programme can reinforce what a teacher is saying. It is by no means an alternative to spoken instruction.
2. There is nothing complicated about a simple two projector show.
3. An audiovisual show can help to vary the pace of a training session.
4. It gives the lecturer a break.
5. At each session everyone gets the same story.
6. Visual images are more interesting if they are carefully organized into sequences and accompanied by synchronized sound.

Training officers and teachers who reject audiovisual may also reject other teaching methods such as audience participation, live demonstrations and so on. They are, though, all essential. An unforgettable lecture should be a fireworks display of ideas, information and techniques.

Multivision, therefore, is used for group training. In its larger form multivision expands to be 'multiscreen' with two or more projectors on each screen area. There is a place for multiscreen shows in training and this is widely used for this purpose in the USA although rarely in Europe. I suspect the reason may have something to do with the relative budgets of universities and training departments on either side of the Atlantic. This is not a direct reflection of economic wealth so much as a measure of the importance given to training by the people who control the budgets. Multi-screen material may be used as a visual aid with push button control, or multiscreen audiovisual shows can be programmed by small encoders. Two or three screens are normally sufficient for showing comparisons. Larger audiences will benefit from having the larger multiscreen formats, although here, as with all projected images, control over the environment in which they are shown is essential.

How effective audiovisual is in training and education is entirely related to how well the material has been organized. Those institutions which can afford specialist audiovisual departments have a head start on the do-it-yourself amateur. But if you are faced with the situation of having to produce your own programmes it is hoped that the other chapters of this book will be of help.

Tourist venue interpretation

In many countries, tourism is one of the chief industries. In fact, in particular towns it may well be the primary money spinner. Flocks of tourists with many different cultural backgrounds will have paid a great deal of money to visit a particular area of artistic, picturesque, historical or archaeological interest. And yet when they arrive and walk around the site they may be disappointed that it does not immediately yield some of its secrets. These secrets may have been discovered and evaluated by experts but the information remains locked within learned volumes, virtually out of reach to the general public.

It is because of this that the need for interpretation was felt. And there is no way more appropriate than installing an audiovisual show for instant on-the-spot interpretation.

Anyone contemplating such an exercise must be aware of the limitations. A slide/sound show can give only the broadest treatment to a subject within the space of a few minutes. The tourist does not want to linger for half-an-hour while dates and details are spelled out. In fact the tourist wants to be entertained, for the simple reason that he is on holiday.

However, there is a different kind of tourist: the visiting student. There may be parties of school children on an educational visit. In this case the show will need to be more information-orientated, and there can be a number of smaller single-screen shows each explaining different aspects of the venue.

The location

Much depends on the site and whether an audiovisual show can be accommodated under ideal conditions. The conditions which are ideal for projection are described later in this book, but there are some special considerations which apply to tourist venues.

First one must consider local regulations and obtain planning permission to convert an existing building or indeed to build a new structure for housing the show. Both the building inspector and the fire and safety officer should be consulted. If no structural work appears to be required, make sure that this is true by consulting a designer who has experience of audiovisual displays.

The *size* of the room, area or structure, is largely dictated by the number of visitors who will be seeing the show at any one sitting. If there is room one should allow at least for a coach load of people (around sixty). This should be worked out in conjunction with details about the *length* of the show and the *frequency* or number of times it is run. It is most important to ensure that each visitor is able to see the show from the beginning and the most successful interpretation programmes are run at set times with a short interval to allow for replacement of the audience by the next group of visitors.

A new building should be designed with the format of the show in mind. An auditorium needs to have a sympathetic *shape* for unusual formats, especially very wide five- and seven-screens 'in a line'. There will also be acoustic considerations which become more important the larger the auditorium.

Furnishing is rarely elaborate at a tourist venue. In practice it is often found that some members of the public can be infinitely destructive. Simple bench-type seating is quite adequate for most shows. An alternative would be rails which keep the audience at a reasonable distance from the screen. A *raked* floor is desirable given that the smallest members of the audience are often the least equipped for fighting their way to the front.

Staffing is an important consideration. Someone needs to be in attendance to start the show, or, if it is totally automatic at least to occasionally check that it is still working properly. The person responsible for this may well be able to sell souvenirs, books and posters and will be on-hand to answer questions.

All of the points I have mentioned imply that the project has a specific budget. In practice, the person who is motivating the venture will have to prepare estimates containing the above details together with costs and submit them to a governing body. The local authority, trust, or branch of government will want to know the degree to which the project is self-financing. A small charge may be levied for seeing the show. Alternatively it can be argued that the audiovisual installation will attract more visitors to the venue.

Financial support may be forthcoming from a tourist board, from the Government itself, or from any one of a number of trusts who can give limited funds towards a carefully planned and worthwhile installation. Alternatively, commercial sponsorship may be attracted, especially if the show gives an opportunity for publicity.

When adequate financing has been achieved, an installation and a programme can be tailor-made to fit both the purpose of the venture and the size of the budget. You have to decide on the choice of a producer and the choice of equipment. Look around first and see what is working properly elsewhere. Visit different production companies, visit manufacturers, but above all, visit other sites and discover what pitfalls to avoid. Remember, too, that audiovisual is still a young medium and that you can contribute to its growth.

The exhibition show

Trade shows and exhibitions are popular occasions for putting on slide/sound displays. Indeed, at some exhibitions, not necessarily those related directly to the audiovisual industry, a visitor from another planet might think that the occasion was a meeting of audiovisual devices, with people (servants) in attendance. At exhibitions one may see the whole range of audiovisual systems chattering away, frequently to each other or to themselves. All of them compete for attention, even such different media as filmstrip and video, super-8 film and slide/sound multiscreen. The ensuing cacophony tends to overload the visitor within a few minutes of his arrival.

What can slide/sound audiovisual achieve at an exhibition? To some extent the answer to this depends upon what the exhibitor hopes to achieve by being represented at the show. Very often firms take part in a trade show simply because they feel that they have to put in an appearance. This, combined with the opportunity to have a few drinks with others in the trade, hardly justifies what can be an enormous expense.

An exhibition stand should be a part of the overall marketing strategy. It can be an occasion for launching a product or a service. It can provide an opportunity to test the market reaction to a product. Or the objective can be simply to obtain as many new sales enquiries as possible. The more objectives you have, the larger your stand has to be.

The audiovisual programme should be an integral part of the stand and should be related to the objectives of it. If the programme is merely 'putting in an appearance', it will do very little good at all. In fact it may even interrupt the business of the stand and confuse the customers. It is far better that you should attach more importance to the audiovisual show, ensure professional standards of production and installation, and you will find that your whole 'presence' at the exhibition is improved.

Specific objectives

Apart from specific marketing objectives, an audiovisual show can perform two roles on an exhibition stand.

It can attract people to the stand. If the show is visible to people passing the stand they will want to stop and watch, if only for a few moments, providing that the show is immediately interesting. To achieve this it will need to be short, say 2 to 4 minutes maximum, fast-moving with a bright rear-projected picture that is shielded from as much incident light as possible. The programme will be on continuous run, probably using duplicate sets of slides in order to cut down resetting time of the projectors.

Alternatively, it can provide a relatively in-depth treatment of a subject to people who, in any case, would have visited the stand. This is the more important role of audiovisual at exhibitions. For a start, the audience is interested in what you do. They are all potential customers. They *are* an audience and should have the facilities which they deserve, namely set times for seeing the show, near-perfect viewing conditions (a mini-theatre within the stand), and protection from the noise of the exhibition. You should certainly consider the provision of seats or benches which will make the proposition of seeing the show even more attractive.

So, with these two roles, we have two distinct types of programme: the first, short, clever, even gimmicky and the second, longer, perhaps 10–15 minutes, and containing more information. Let us now briefly look at the staging of each type.

The attention-getting show

The number of screens you use depends on the show, and your stand. Because of the need for speed, I should recommend using three or more projectors on each screen. This allows for faster changes of image. Individual screen sizes should be kept small to produce a bright picture. Unusual formats are attractive in this situation, particularly in multiscreens. The show needs colour, pace and entertainment value.

All of these points are, for once, *more* important than what the show actually says. The objective is to get people to linger on the stand just long enough for one of your staff to 'buttonhole' them. Better still, the programme should raise a few questions so that the visitor himself approaches one of the staff.

The equipment which is used for this type of show should not require adjusting after it has been switched on at the beginning of the day. It should work on a continuous run basis and it will give your stand the appearance of being busy, even during quieter periods. The audio volume should not be so high as to inconvenience other exhibitors. Use commercial-type loudspeakers for a crisp sound reproduction rather than hi-fi cabinet loudspeakers which have a strong bass response.

The near-projected picture can cause a few design problems. A fairly long projector throw is desirable in any audiovisual installation. Short throw lenses cause loss of registration because two or more projectors are being used. They cannot share the same optical axis. However, exhibition space is expensive and you may be reluctant to use much of it (say 10–12ft) for this purpose. An alternative is to use mirrors or a specially designed rear-projection cabinet that has a mirror system built into it.

A similar conflict arises with the attempts to shield the screen from incident light. The more you shield it, the less visible it is from a wide area of the exhibition. And the chief role of this particular type of show is to attract attention.

Elegant solutions have been found to both of these conflicts by careful design. Figure X shows one of the installations at the Silver Jubilee Exhibition in Hyde Park in 1977. The show had a higher information content than those of the type I have been discussing, but it was able to play a more important role because of the design features of the exhibit. Using hexagonal screens enabled registration to be more easily achieved since the corners of each frame had been masked. Seven screens were used which enabled the individual size of each to be limited. The whole screen area was protected in a type of geodesic dome which allowed people to see that a programme was running, even from different parts of the exhibition. The programme was produced by Prater Audio Visual of Greenwich for Lloyds of London.

Less prestigious installations may (and should) have equally elegant design solutions. If front projection has to be used, then the screen must be recessed towards the back of the stand and the projectors and control equipment hidden within an enclosed projection box.

The informational show

When the presentation is intended to play a major part in the exhibition, it needs its own theatre. Here is an opportunity for both the sponsor and the designer to be adventurous. While the first solution might be a conventional shape of theatre, that is, an oblong shape with a flat screen and rows of seats, there are many other formats to choose from. For example, there is projection 'in-the-round', a connecting row of screens running along the walls of a round theatre. The audience can enter the area via a doorway beneath the screen, and exit the other side. The projectors are recessed into the walls behind glass, preferably below the screens which should be tilted slightly downwards.

Gimmickry for the sake of it should always be avoided although an exhibition does require the participants to be 'exhibitionists' as well as being 'exhibitors'. It's no use keeping a low profile, quietly carrying on regular business because you can do that anyway. A public display is an opportunity for a celebration, and a good multivision show ensures that visitors to your stand will at least remember whatever it is you are celebrating.

A good example of the informational multivision show was to be seen at a recent Daily Mail Ideal Home Exhibition in London. The particular stand was named 'The TV Times House of the Stars' and was in fact one of the most ambitious projects of its kind ever constructed at this regular annual show. It consisted of five interlocking rooms, each decorated according to the choice of a well-known personality. Twenty-five sponsors, mainly furnishing manufacturers, took part and supplied the contents of the rooms. Each room had a small multivision show explaining the choice of furniture and showing details about it. In all, there were three two-screen shows, two single-screen shows and one three-screen.

All the projectors were individually controlled by plug-on control units, and were centrally mounted on a substantial tower, built in the 'core' of the stand. Five NAB standard cartridge decks carried sound and multiplex control tracks. One of the main features of the installation was the control of spotlights. Each room had sixteen channels of lighting effects, that is, up to eighty circuits of lighting could be individually switched on and off at any point in the show. In this way items of furniture, even small details, could be highlighted in order to draw the attention of the audience to them. The shows, which ran continuously throughout the four week exhibition were made by Peter Hirst Smith Photography, the stand was designed by Chris Miles of Cairns Maltby and the audiovisual and lighting control was specified by the author on behalf of Electrosonic Ltd.

3   Projecting the image

The basic building block of any multivision show is the automatic slide projector. Without this invaluable means of projection there would be no audiovisual, no multiscreen spectacular, no single-screen show. In particular, it is the automatic slide projector with a circular magazine of slides which is a universal standard throughout the industry.

In Europe, the Kodak Carousel S-AV 2000 has become the most widely used projector in multivision shows. In the USA the Kodak Ektagraphic is employed, although the more sturdy S-AV 2000 is sometimes specially imported for use in multivision.

The S-AV 2000

The Kodak Carousel S-AV 2000 has evolved over a number of years, the principal aim of the manufacturers being to produce a heavy duty machine that will cope with continuous running at exhibitions and publicity applications. The latest version is the S-AV 2000 G. It has a slide capacity of 80 which are gravity-fed into an accurately registering slide gate.

The slide tray has a transparent cover which keeps slides from falling out when the tray is removed from the projector and also keeps them clean and relatively free from dust.

Each slide tray has a metal base plate that holds the slides into the tray. A slot allows a single slide at a time to drop into the gate. The tray rotates and it must engage a latch in the zero position before placing on the projector. If it has not engaged at zero, turn the tray upside down (with cover locked in position) and rotate metal base plate.

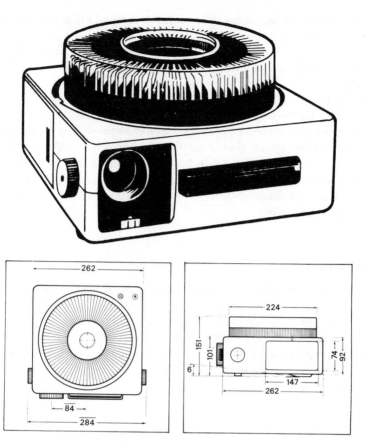

Kodak Carousel S-AV 2000 automatic slide projector. Dimensions in mm of S-AV 2000 projector.

The S-AV 2000 uses any 5×5 cm (2×2 in) slide with a maximum thickness of 3·2 mm ($\frac{1}{8}$ in). It is designed to the standard DW 108.

The projector is provided with a remote focus facility while this may be useful to a lecturer operating a single projector, it is not used in multivision shows. In any case, it is preferable to prefocus the projector and use glass-mounted slides to retain exact focus throughout the show.

Power supply

The projector can be used on alternating current between 110 and 250 volts, 50 or 60 Hz. There is a voltage selector on the underside of the projector which should be set to the appropriate voltage, using a screwdriver or a

33

coin to turn it. When using other control equipment, make sure that the voltage is also selected on these as well.

The mains fuse is located next to the voltage selector. To change it, unscrew the cover and pull out. Different fuses are used for the two voltage standards.

For 220–250 volts (e.g. UK) a 1·25 amp slow blow is fitted.

For 110–130 volts (e.g. USA) a 2·5 amp fuse is fitted.

The projector should be placed on a firm support to give vibration-free projection. This support should normally be horizontal, although the manufacturers recommend that it can be tilted by up to 30° to the horizontal in any one direction without affecting the gravity-feed operation.

At the sides of the projector, two black milled knobs allow for adjustment of height and levelling.

Ventilation

The projector should have sufficient space around it for drawing in cool air via the lens carrier, gate aperture, and vents in the side of the lamp-house door, and propelling warm air away at the back.

An automatic thermal cut-out will operate if the projector becomes too hot. This occurs only if, for instance, a slide has jammed in the gate, and not with normal continuous operation. When the projector has cooled it switches on again.

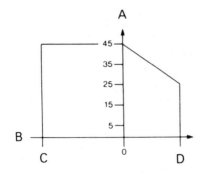

E		B	
240/250	212	250	275
220/230	195	230	253
130	115	135	148
110	98	115	126
F	20.4	24	26.4

Operating range of Carousel S-AV 2000 projector. This is the permissible operating range in terms of room temperature and mains voltage. A: Room temperature (°C); B: Mains voltage (V); C: 15% under voltage; D: 10% over voltage; E: Voltage selector position; F: Lamp voltage (V).

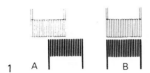

1   A   B

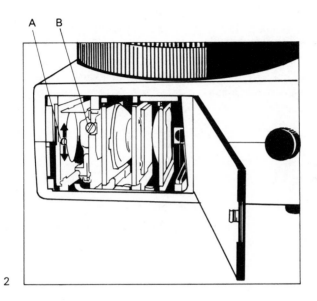

A   B

2

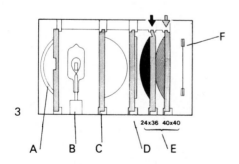

3

24x36   40x40

A'   B   C   D   E

Optical system of S-AV 2000 projector.
1, Centering filament and image for exact alignment of lamp. A: Not centered; B: Correct. 2, Access to optics and lamphouse. A: Mirror alignment lock; B: Lamp alignment screw. 3, A: Mirror; B: Lamp; C: Condenser; D: Heat-resistant glass; E: Positions of second condenser.

Lamps

The S-AV 2000 is fitted with a tungsten-halogen lamp (24 volts/250 watts). This runs at a standard or economy setting. The economy setting is normally used for rehearsal purposes. When using a single projector, the settings are obtained by inserting the appropriate side plug into the 12-pin socket at the side of the projector. A single projector *must* have a side plug fitted or the lamp will not operate.

The projector should always be switched off when fitting a side plug because the pins can be burnt and contacts fused together.

The following table shows approximate figures for lamp life for a 250 watt lamp and a nominal 24 volt supply. These depend on the individual make of the lamp (and here are related to cost), and they vary according to whether the lamp is over- or underrun.

Setting	Average lamp life (hours)	Average light level (%)
Standard	approx 50	100
Economy	approx 200	70

In a large multivision installation it may be necessary to change lamps during a show. This is an emergency measure and should not be attempted by anyone untrained in the operation. New projectors may be fitted with an automatic lamp change unit to obviate the need.

A projector lamp gets very hot in operation and it is preferable to allow it to cool before removing. Disconnect the plug, open door to the lamphouse, pull adjustment screw to swing lamp down to an accessible position and remove, using gloves or a cloth to protect your hand.

When replacing a new 24 volt/250 watt tungsten-halogen lamp it is very important not to get finger marks on it. This impairs the light transmission and can even shorten the life of the lamp. Keep the lamp in its cardboard sleeve until it is pressed firmly home, then remove sleeve. Then swing the lamp holder back into position.

The adjustable screw is used for exactly aligning the lamp with the optical system of the projector. A fixed reflector returns the light along the condenser lenses that supply even illumination to the slide, towards the focusing lens. It is therefore essential that the position of the lamp is exactly centred according to this system.

With *zoom lenses* and *60–100 mm* lenses, the lamp filament may be observed through the lens itself—with the lamp switched off. Two filament images will be seen.

The two images should be as close to each other as possible, but not touching. Their separation can be adjusted by moving the stand on the reflector panel up or down.

With long focal length lenses, e.g. 150 mm, 180 mm and 250 mm, a pinhole slide should be placed in the slide gate. A plastic lens cap should be fitted and with the projector lit the lamp filament images can be seen projected on to the lens cap. The lamp is centred as before.

With wide-angle lenses, e.g. 28 mm or 35 mm, the projector should be pointed at a white wall about 1 metre away, without a slide in the gate. With the lamp on, the lens should be unscrewed until the lamp filament images are visible on the screen. Centre as before.

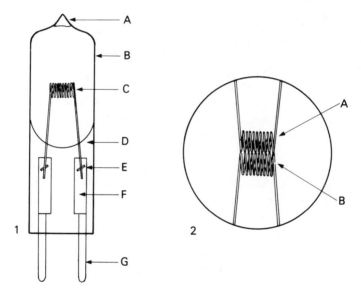

1, Low voltage, tungsten-halogen projector lamp. A: Exhaust tube seal; B: Fused quartz bulb; C: Tungsten filament; D: Quartz pinch; E: Platinum weld; F: Molybdenum foil; G: Molybdenum pins. 2, Rectangular filament and image forms a square and more effective light source. A: Filament image; B: Filament.

Condensers

The Kodak Carousel S-AV 2000 is provided with interchangeable condensers which may be used to match the lens being used. The condenser system is designed to provide even brightness over the whole image that is projected. Standard condensers are used with slides up to 24×36mm and lenses of 60 to 100mm (and zoom lenses).

When using 38mm $\times$ 38mm superslides the condenser nearest the slide should be placed in the last slot on the right.

With long focal length lenses a special condenser lens should be used. This replaces the condenser closest to the slide gate.

With each of the wide-angle lenses, 28mm and 35mm, Kodak provide special condenser lenses to supplement the system. With the 28mm lens the special condenser should be fitted closest to the slide gate. In this case both condensers must be slotted in together because of the close spring.

When using wide-angle lenses you should remove the black lens diaphram from the front of the projector, otherwise correct focusing is not possible.

Older carousels

It is worth mentioning that a number of models preceded the S-AV 2000 G. Many of these are still operational and may be encountered from time to time. Here is a list of models up to the time of writing.
1. 'S' TYPE   Two-tone, grey/white; 150 watt lamp; May have plastic front condenser; 12-pin DIN side socket.
2. S-AV   Grey; Early models similar to 'S' type (150 watt lamp and no side socket); Later models 250 watt with 12-pin DIN; Metal 12-pin socket surround; Round lens holder aperture which does not allow 35mm lenses (or wider) to be used.
3. S-AV 28   Similar to later model SA-V, but with square lens aperture allowing 28mm (for 26×26mm slides) and 35mm (for 24×36mm slides) lenses to be fitted.
4. S-AV 2000   250 watt only; 12-pin DIN with plastic surround; Square lens holder aperture.
5. S-AV 2000 G   Metal tray bearing; Larger plastic surround on 12-pin socket; Several internal differences; Different snap solenoid kit required.

Slide changing

Control of slide changes can be effected from the projector itself. Two black buttons on the top of the projector step the projector forwards and backwards. The 'forwards' button also allows the slide tray to be rotated manually to any given slide if it is held down.

To change slide tray, always rotate it to position 0 before attempting to

1, Always use blank slides in zero position. Insert before attaching tray. 2, Attach lens support bracket for long focal length lenses.

remove it. It is a common mistake to disconnect the projector before re-moving tray. Remember, you need the mains power in order to get back to zero!

Remote control

The functions of the forwards and backwards buttons can be operated by remote control up to a distance of 20m (66ft). Beyond this distance, the voltage drop in the remote cable tends to give unreliable operation.

The remote control is plugged into a 6-pin DIN socket on the rear of the projector. If auxiliary control, i.e. a dissolve unit, is used, then the remote control is normally plugged direct into the unit and not into an individual projector.

An interval timer may be used with a single projector. This advances the projector at preset intervals between 4 and 30 seconds. It plugs into the remote control socket.

A tape recorder may also be used to advance a single projector using pins 2 and 3 of the remote control socket for forward slide changes. There are a number of tape recorders on the market with built-in synchronizers. It is possible, although I should not recommend it, to use a twin socket adaptor in order to operate remote focusing with tape recorder control.

For other accessories there is a DC voltage of 20 volts (maximum current 200 amps) across pins 3 and 6 of the 6-pin socket.

It is the 12-pin side socket of the S-AV 2000 which allows the sophisticated control that can be obtained by the wide range of units that are available.

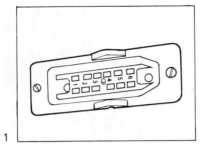

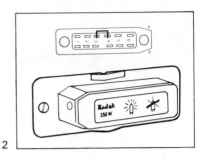

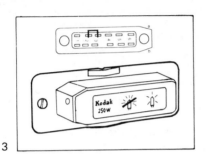

1, 12-pin socket to DIN 41622. 2, Standard lamp setting. 3, Economy lamp setting.

The connection with a control unit is made with 12-pin knife-contact plug to standard DIN 41622.

The slide changing and lamp setting are separately controlled, so the slide may be changed when the lamp is switched off and the lamp itself may be independently controlled by an external unit. The lamp circuit is of low resistance and in order to prevent loss this must be duplicated by the external unit. Cable, relay and contacts should be of adequate rating and the resistance of the wiring must not exceed 0·1 ohm.

The Kodak Ektagraphic projector

The automatic slide projector used in many multivision systems in North America is the Kodak Ektagraphic projector. The Ektagraphic is similar to the S-AV 2000, having a circular slide tray and automatic forwards and reverse stepping. It does not have automatic zero position detection. This needs to be fitted for autopresent, that is, completely automatic operation. Other major differences between the Ektagraphic and the Carousel are in the light source and optical system.

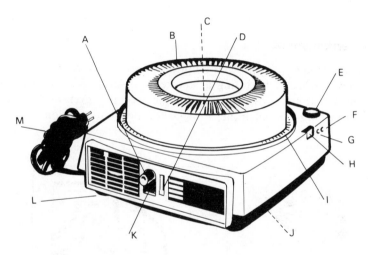

The Kodak Ektagraphic Automatic Slide Projector. A: Remote control receptacle; B: Tray; C: Emergency tray removal screw; D: On/off, high/low switch; E: Focus knob; F: Forward; G: Reverse; H: Select; I: Gate index; J: Lamp door (on bottom); K: Dissolve control receptacle; L: Levelling foot; M: Power cord.

Projection lamp

The standard projection lamp for the Ektagraphic is a 300 watt, 120 volt ANSI Code ELH lamp. Light output is equivalent to that from the Kodak Carousel. Increased wattage and voltage does not necessarily mean a higher light output.

A choice of lamps is available for the Ektagraphic.

ANSI Code	Relative brightness		Average life (hours)
ELH (medium brightness/life)	Low	70	105
	High	100	35
ENH (maximum lamp economy)	Low	50	330
	High	65	175
ENG (high brightness)	Low	90	50
	High	130	15

Lamp replacement

To replace lamps, turn projector upside down and open the lamp door. The lamp is released by lifting a lamp ejector lever. When replacing the lamp check that it is correctly centred in the two-pin socket. Push the socket towards the lamp and latch the lamp ejector lever into position.

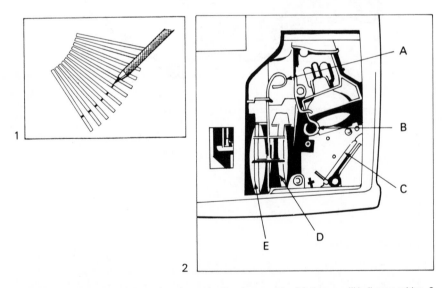

1, If both front and rear projection is to be used, a line drawn with a felt tip pen will indicate position. 2, Inside lamphouse of Ektagraphic projector. A: Lens retainer; B: Lamp ejector lever; C: Mirror; D: Heat-absorbing glass; E: Condenser lens (flatter side away from lamp).

Optical system

The Ektagraphic projector has a disc of heat absorbing glass and a condenser lens both of which are clipped firmly in position by a lens retaining clip.

Heat-absorbing glass should be handled with care, since, during manufacture, it undergoes special processes which put stresses and strains into it. This glass easily shatters for no apparent reasons. The manufacturer recommends the following precautions:

Use a glove or a piece of cloth when handling the glass.

When the glass is removed place it on a piece of insulating material such as wood, rubber or cardboard.

The glass should be covered while it is removed.

The reflecting mirror should not be removed because it has been carefully aligned during manufacture to provide the best illumination. The mirror surface should not be touched with the fingers. Both sides of the heat-absorbing glass and the reflecting mirror may be cleaned with lens cleaning paper.

Controls

The Ektagraphic projector is controlled by external equipment via a 7-pin connector which plugs into the dissolve control receptacle at the rear of the projector. The projector may be manually operated by remote control by connecting a hand controller to the 5-pin receptacle.

Other features include a selector switch for turning on the fan (independently) and the lamp to either low (70%) or full brightness. Manual focusing is achieved by turning a focus knob on the top of the projector rather than by twisting the lens itself.

Slide trays

Carousel Universal and Carousel 80 slide trays will fit all types of Carousel or Ektagraphic projectors. An alternative tray with increased capacity is the Carousel 140 tray that holds 140 slides. This is not recommended for multivision use because it will only hold slides with a maximum thickness of 1·6mm. Glass-mounted registration slides require 80-slide capacity Universal trays.

	Maximum thickness of slides
Carousel Universal	1/8 in, 3·2 mm
Carousel 80	1/10 in, 2·5 mm
Carousel 140	1/16 in, 1·6 mm

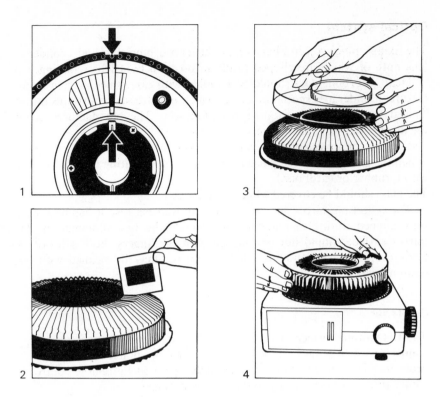

1, Underside of circular slide tray. Arrows show gate position of metal base plate. 2, Loading slides into tray. For front projection slides should be upsidedown and 'right-way-round'. For rear projection slides should be upside-down and 'wrong-way-round'. 3, To secure cover, lock with clockwise twist. 4, Loaded tray can be placed on projector only at zero position.

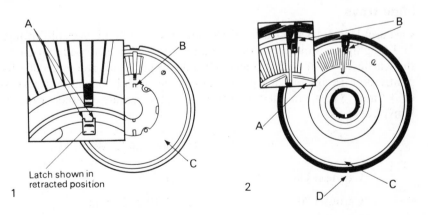

1, Carousel Universal and Carousel 180 slide tray. A: Latch notches; B: Latch; C: Slide retainer plate.
2, Carousel 140 slide tray. A: Release; B: Latch; C: Index hole; D: Index notch.

44

All slide trays consist of a metal slide retainer plate (or base plate), a method of latching this plate at the zero position, and a locking ring for keeping the slides firmly in their compartments. Slides are numbered 1–80, or, on the 140 tray only the even numbers are marked.

'Transvue' trays are available from Kodak which will allow you to see the slide identification numbers in total room darkness when the projector lamp is on. These are made of a translucent material that allows the projector light to illuminate the numbers.

How the automatic slide projector works

As an example, I have chosen the S-AV 2000 G. However, other automatic projectors work on similar principles.

Mains power is attached via a three-pole input socket. The live is switched, fused and routed via thermal cut-out and voltage selector to the mains transformer.

The mains transformer primary has three sections, connected in different configurations by the voltage selector; to give a choice of 110, 130, 220/230 and 240/250 volts. Neutral is applied to the other side of primary wiring.

The thermal cut-out is mechanically connected to lamp plate. Early models have a reset facility. New models automatically return power when projector has cooled. Operation of cut-out has the same effect as switching the projector off.

A 125 amp 20mm slow blow fuse protects mains circuit against short circuit. A short circuit on 24 volts secondary will not necessarily blow mains fuse.

The fan motor has two 110 volt windings, which are connected across 110 volt windings of transformer. Like the transformer, the fan motor is controlled by the on/off switch, thermal cut-out and fuse.

The secondary winding of transformer has three connections: Common; 24 volt AC; and 22 volt AC. These are brought out to the 12-pin DIN side socket. One side of lamp is connected to 24 volt common, internal connection. The other side is connected to 12-pin socket. Therefore, by an external connection (side plug or control unit), either the 24 volts or 22 volts will operate, giving either full brightness or economy lamp setting. Lamp current is 10 amps.

The 24 volt supply is used to operate the solenoids and focus motor. This supply is connected to a bridge rectifier which changes the AC current to DC. The common is also connected to the bridge rectifier.

The fan motor runs while mains is switched on. The fan draws cooling air into the projector via the lens carrier, the gate aperture and the vents in the side of the lamp house door. All these routes are important for the cooling operation and the air is directed over the slide in the gate, the condensing lenses and the lamp.

The mechanical functions of the projector, that is, magazine forward/reverse stepping; shutter opening/closing; gate opening/closing; and slide raising/lowering are controlled by the motion of four large cams. These cams are operated by the fan motor to which they are connected by a clutch mechanism. The clutch is solenoid operated.

The reverse solenoid and microswitch, the forward and reverse buttons (these are in fact microswitches) and the remaining two microswitches which are operated by the cams complete the electrical control system of the projector.

Common faults and cures

There are a number of common faults which may occur, as one would expect, since a projector is an electromechanical device which is subjected to extremes of wear and tear.

Blown fuse The fuse may blow for a number or reasons, or for no apparent reason. The cause is likely to be an electrical short circuit if the fuse itself is blackened. But if the wire is simply broken, replacing the mains fuse may restore the projector to perfect working order.

Blown bridge This causes the forward and reverse buttons to be out of action. The bridge rectifier card (four diodes) should be replaced or repaired. Bridge failure can be caused by earth faults (pinched wires), external control unit faults or internal component failure.

Bridges do fail, in spite of precautions by the manufacturer, because of the induced voltage spike when projector is switched off. The voltage dependent resistor (VDR) connected across the transformer primary in G-type projectors has helped to prevent this.

Jamming Everyone who has used projectors frequently will have encountered a jamming of the slide magazine. The main cause is insufficient slide lift height. There is a screw adjustment on the lift arm which should allow for a slide lift of 3·5 mm.

This can be checked by rotating a full slide tray and ensuring that there is a pronounced ripple effect where the slides cross the tope of the lift arm.

The aperture of magazine base plate must be flat. If the edges are bent upwards the slide needs to be lifted higher than by the normal amount. All new magazines should be checked.

A second cause of jamming can be an incorrect adjustment of the reversal lever. The amount of stroke between forward and reverse steps should be equalized.

A third cause can be when the tray transport lever does not disengage from between the pegs of the magazine. This may be because of a weak transport lever spring or a build up of dirt.

Lamp not lighting The most obvious cause is a blown lamp. Replace. It can however be (a) a burnt lamp holder (b) burnt connecting crimps (under voltage selector) or (c) burnt pins on side socket.

Slides not dropping correctly (a) Gate may not be opening fully. Needs lubricating? (b) If slides only drop half-way, this is probably due to the disconnection of the small gate lever which operates the lateral slide clamp. (c) Gate lever may be distorted. (d) Magnetic snap solenoid (if fitted) may be attracting slide lift arm. A slight 'dog-leg' may be put in the arm to increase the distance from the solenoid.

Long run applications

Multivision installations that give a number of showings each day (and there are many which run continuously throughout the day) require the projectors to give trouble-free operation for long periods. It is essential to provide adequate ventilation to and from the openings in the projector housing. In some cases it will be necessary to provide forced-air ventilation. Normal room temperature is adequate, but it should be dust-free.

Servicing

Finally, all projectors should be serviced regularly. A complete service should be given after every 500 hours of operation. Modern methods are available, including ultrasonic cleaning in which the projector is dipped in a 'dry-cleaning' bath and dirt is vibrated out by sound waves. This removes every particle of dirt, even from the most inaccessible places, and mechanical parts need to be lubricated afterwards.

High power projection

In the staging of audiovisual shows there are many situations where high intensity projection is essential. In particular, when the following conditions apply:

Front projection, with an extra long projection throw.

Front or rear projection, with an extra large image size.

Rear projection, with high ambient light.

In these situations it is essential to use equipment which will provide a high intensity lamp. There are, in fact, a number of xenon arc projectors on the market. They are converted Kodak projectors with a circular slide tray to carry out the mechanical functions of slide handling.

Basically, there are three types of xenon arc projector available:

Single projector, manual and timer operation only.

Dual projector, with single light source, manual and timer operation only.

Single projector with fade module, fully usable for tape and manual control.

The lamp intensity of xenon slide projectors will vary from model to model, from 4000 lumens to a maximum of 5500 lumens. This is the brightest light which can be forced through a normal transparency without damage. It is about four times the brightness of an ordinary slide projector. All xenon arc projectors are rated at a colour temperature 5400°K, that is, daylight temperature. Slides need to be colour balanced especially for xenon projection.

Lamp modules are expensive to replace, costing more than an ordinary quartz halogen projector. They will, however, provide an average of 1000 hours of operation, and will normally be guaranteed by the manufacturer for around 300 hours. Operating costs should not exceed those for a quartz halogen projector. Light from the lamphouse is reflected by a non-colour sensitive metal reflector to ensure that the colour temperature is identical for each projector. This enables a user to show a number of images side by side on a multivision screen without any variance in either colour temperature or intensity. The condenser optics spread the light evenly across the slide.

During operation of xenon arc projectors care must be taken to allow the cooling fans to operate both before lighting the lamp and after switching it off.

Unfortunately, the weight and bulk of xenon projectors prevents them from being used more frequently for travelling presentations. The power supply represents 50% of the total weight but this is necessary for supplying the increased load: normally 10 amps for a single projector, 20 amps for a dual projector with manual dissolve. Lamp house, power supply and mechanical projector are normally detachable and may travel separately.

Fade/dissolve xenon projectors

A xenon light source may be controlled by a potentiometer in the power supply, but this is used only to vary the output at the upper end of the brightness range. In order to dim the light down to total darkness mechanical means are used. A servo-operated disc is used to progressively prevent light

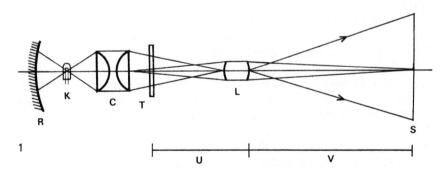

1, Condenser system images the lamp filament to fill the entrance pupil of the projector lens. R: Reflector; K: Lamp; C: Condenser; G: Gate; T: Transparency; L: Lens; S: Screen; U: Object conjugate; V: Image conjugate.

from reaching the screen. This may be remotely controlled by a potentio-meter and hence by a normal dissolve unit.

In this way it is possible to control two single xenon projectors fitted with fade modules in exactly the same way as ordinary Kodak Carousel or Ektagraphic projectors. Each xenon arc projector can be fitted with an individual control unit and operated from a memory programmer.

The major application for large numbers of xenon slide projectors is in the theatre, particularly in the spectacular multimedia musicals which use a host of visual effects. For instance, '*Beatlemania*', the highly successful Broadway hit about the 1960s, uses fifteen xenon projectors to provide projected images on screens which are close to areas of the spot-lit stage.

Projection lenses

The problems encountered in designing optical systems for projectors differ from those in camera lens design. A camera lens images a three-dimensional subject, which may be a great distance away, on to a flat plane situated a few centimetres behind it. If the image were replaced by a slide transparency and a light source positioned behind this, then the optical system would be totally inadequate for slide projection. Camera optics will not work in reverse.

There are three main differences:

There is now an artificial light source that must itself be made to be directional. A condenser system is required.

In the reverse optics of projection the object conjugate is now many times less than the image conjugate.

Not only the image (the screen) but the object (the transparency) is a flat plane. The amount of light passing through the lens needs to be maxi-mized rather than controlled by a variable aperture.

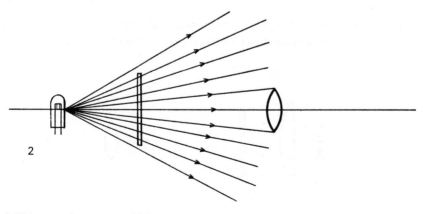

2

2, Without condenser—wasted light.

Basic projection optics

A basic projector system consists of a light source, a condenser lens system to direct light through the transparency and into the entrance pupil of the lens, an imaging lens, and a screen.

Condenser systems

Any loss of light at the entrance to the imaging lens must be minimized and therefore the condenser system will be corrected for spherical aberration and may have anti-reflective coatings. The lamp filament is focused to a sharp image within the imaging lens by the condenser system. The condenser system is therefore critical and on many projectors is adjustable so that it will match the type of projector lens used. The light from the lamp should exactly fill the rear element of the projector lens so that light from the edges of the transparency is transmitted to the centre of the element and light from the centre of the slide is transmitted to the outer edge of the entrance pupil.

The transmission of infra-red energy must be kept to an absolute minimum in order to prevent heat damage to the slide. This may be done in a number of ways.

The most important is the incorporation of a heat filter which may either absorb or reflect infra-red radiation.

The reflector can be coated to transmit infra-red radiation and reflect visible light. Infra-red reflective coating can also be placed on the sides of the condenser elements facing the light source.

Mechanical cooling by fan is essential, and mounting the slide in a white mount helps to reflect heat, and protect the slide mount itself.

A combination of these precautions ensures an even illumination without intense heat.

Condenser lenses have large diameters in order to collect the maximum amount of light. Because of this, two or more elements are required to produce

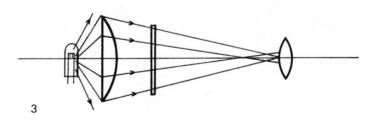

3

3, With condenser—more light.

the short focal length which is necessary in projector construction. The number of condenser elements can be reduced by giving the rear element an aspherical surface. This is done by moulding and flame polishing. The aspherical surface faces away from the lamp source while the element itself will be positioned as close as possible to the lamp.

Projector lenses

The imaging lens of a projector needs to have a large aperture to give the maximum illumination on the screen. This aperture is always fixed since there is no need to control the amount of light passing through it, unlike the variable aperture of a camera lens that has to be used with light-sensitive emulsions.

The following table shows a selection of projector lenses that at the time of writing are available.

Focal length (mm)	Focal length (in)	Aperture	Manufacturer
15	0·59	$f2·8$	Aga
25	0·98	$f2·8$	Berthiot
35	1·37	$f2·8$	Berthiot, Kodak
50	1·96	$f2·8$	Berthiot
60	2·36	$f2·8$	Kodak
70–120	2·75–4·72	zoom	Kodak
80–125	3·14–4·92	$f2·9$ zoom	Berthiot
85	3·34	$f3·5$	Kodak
100	3·93	$f2·4$	Berthiot
110	4·43	$f2·4$	Berthiot
115	4·52	$f2·4$	Berthiot
150	5·9	$f3·5$	Berthiot, Kodak
180	7.08	$f3·5$	Kodak
180	7·08	$f4·0$	Berthiot
210	8·26	$f4·5$	Berthiot
250	9·84	$f4·8$	Berthiot
290	11·14	$f4·8$	Berthiot
360	14·17	$f4·8$	Berthiot

A wide variety of lenses is available for projecting 50×50mm slides, ranging from 15mm up to 360mm. Effectively, the practical range for use in multivision systems of two or more projectors lies between 25 mm and 250 mm. Zoom lenses are also widely used. These are not normally true zoom lenses which retain focus throughout their range, but are *varifocal*; so they have to be refocused after each change of focal length.

The lenses of longer focal length necessarily absorb more light and consequently give a less bright image on the screen. In practice this is compensated by the fact that the image itself is smaller, although the size of the image is also related to projection distance and object size (slide format).

Calculating projection distances

A simple formula may be used to calculate the projection distance that is required, or, changed round, the formula will tell you what focal length lens should be used:

$$\text{Projection distance} = \frac{\text{picture width} \times \text{focal length of lens}}{\text{slide aperture}}$$

The projection distance is measured from the centre of the screen surface to the centre of the lens (in practice, the front plate of the projector). This is only a practical guide. The distance, theoretically, should be measured from the second nodal plane plane of the lens. In optics this is the plane from which the focal length is measured. Projection distance should be expressed in the same terms as slide aperture (mm). For example, if your screen is 2 m (6 ft) across and you have a 210 mm lens, using standard 35 mm landscape slides, then:

$$\text{Projection distance (in m)} = \frac{2 \times 210}{35}$$
$$= 12 \text{ m}$$

Calculating correct size of lens

Alternatively, to choose the lens of correct focal length the formula can be changed to read:

$$\text{Focal length of lens} = \frac{\text{projection distance} \times \text{slide aperture}}{\text{picture width}}$$

If you have a screen which is 8 m (25 ft) away from the projector and this screen is 1·6 m (5 ft) wide, then with 35 mm landscape slides the focal length of lens required can be worked out as:

$$\text{Focal length of lens} = \frac{8 \times 35}{1·6}$$
$$= 175 \text{ mm (nearest available lens size, 180 mm)}$$

Calculating picture width and height

The formula may be used to calculate either the height or the width of the picture, providing that the corresponding measurement of the slide aperture is used.

$$\text{Picture width} = \frac{\text{projection distance} \times \text{slide aperture (width)}}{\text{focal length of lens}}$$

$$\text{Picture height} = \frac{\text{projection distance} \times \text{slide aperture (height)}}{\text{focal length of lens}}$$

It may sometimes be necessary to calculate both picture height and picture width if a mirror system is being designed. In this case the light path may be drawn out in either plane i.e. in elevation or in plan. An elevation or vertical cross section of the mirror system will show the height of the image.

Slide apertures

The standard slide formats are $24 \times 36\,\text{mm}$ (standard 35 mm) and $38 \times 38\,\text{mm}$ superslide.

In working out the above equations it must be recognized that there will be small variations according to the type of slide mount used. But for all practical purposes the actual part of the slide which is projected will be slightly smaller due to the masking of the slide mount. This is necessary to give a clear, hard edge to the projected image. For 24×36 slides the measurements to use in calculations are $23 \times 35\,\text{mm}$. With superslides use $38 \times 38\,\text{mm}$.

Types of lenses

As camera lenses can be divided into approximate categories according to their focal length, so can lenses for projectors. However, a standard lens for a projector is approximately twice the focal length of a standard camera lens. The following categories apply for projector lenses.

		price category
Ultra wide-angle	15 mm (fish-eye)	c
Wide-angle	25 mm, 28 mm, 35 mm	b
Standard	85 mm, 100 mm	a
Long focal length	180 mm, 250 mm and upwards	b

Standard lenses of very high quality are inexpensive compared to those in other categories. An approximate guide shows the comparative prices for wide-angle and long focal length lenses to be around six or seven times that for a standard lens. Ultra wide-angles are particularly costly, around 40 times the price of standard lenses. The reasons for this are that not only are they more expensive to make but they are made in much smaller quantities.

53

Lenses may also be categorized according to their design features. We have already mentioned zoom or varifocal lenses. A typical varifocal lens has a number of elements which are moved mechanically between the fixed elements.

It was not until the 1960s that good zoom lenses were available, However, you can still not expect the same quality and minimum of aberration which is obtainable from the best fixed focal length lenses.

Projector mounting

Each projector, and associated control gear, must be mounted firmly. Most audiovisual manufacturers can supply suitable frames for commonly used projectors. The projectors must be as close to the central axis of the screen as possible. The normal way is to mount them in pairs one above the other. Autopresent units quite often hold two projectors side by side on top. When setting up a group of projectors to cover one screen, ensure that their lenses are as close together as possible, and are evenly spaced around the screen axis.

It is quite simple with a piece of paper to find the mean optical axis of a group of projectors. Just draw lines connecting the centres of each distant pair of lenses. With four projectors, this gives a single point. With more

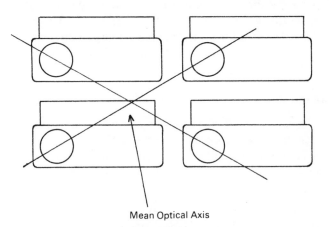

Mean Optical Axis

A far projector set-up to show where the mean optical axis falls.

projectors, this can produce a number of points. In which case, the mean optical axis is midway between them. You then can make sure that the mean optical axis coincides with the screen axis; and thus that you have minimised the angle at which each projector reaches the screen.

When projectors are aligned on a screen from different viewpoints, each image is slightly distorted. The effect, often called keystone distortion is produced because the projector beams are not shining exactly at right angles to the screen surface. There are a number of systems available for minimizing the interlens distance. They may prove useful. Naturally the closer the screen, the greater the distortion. So, in many cases, choosing long-

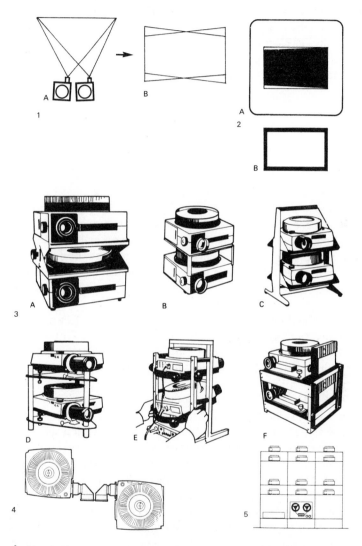

Projector stacking. 1, Short throw projectors side by side, A: result; B: distortion of images (exaggerated in diagram). 2, Solution A: use of offset mounts; B: mask off distortion on screen. 3, Solution: twin lens projectors one above the other. This is the generally accepted solution. Dozens of different kinds of twinning stands are available; A, B, C, D, E, F. More sophisticated stands will have adjustments for aligning images on the screen, and for retaining alignment. 4, Solution: reversing lenses may be used to bring the optical centres close together while keeping the projectors in the same horizontal plane. 5, Solution: tiered shelving may be specially constructed for holding a whole bank of projectors.

focus projection lenses and a long throw is the best way to minimize the problem. A single screen can be given a matte black surround, so that the projector image fills the white area and the distortion is lost onto the black. Such an approach is not always practical in multiscreen presentations. The borders between separate image areas can be obtrusive.

If the image has to be exactly rectangular, the answer is to use tapered slide mounts. Obviously, these have to be tailored to your particular prooejection set up. The wider edge of the slide-mount goes nearest the optical axis. Naturally, you can distort the frame both vertically and horizontally, but the calculations become rather tedious.

Screens

The screen material and position you choose are dictated by the viewing conditions. On areas of high ambient light, you need a highly reflective screen or rear projection, as discussed in the next chapter.

Beaded or lenticular screens, or special curved metallized ones, give the brightest image, but your audience must be close to the projector-screen axis to benefit. That is because most of the light is reflected more or less straight back. Once you move to the sides, the image becomes much less bright.

This directional property can be used to reduce the effect of other lights. With a highly directional screen, lights from the side have little effect on the picture. So you can maintain a high contrast in quite bright ambient lighting.

Matt white screens allow a much wider angle of view, and tend to show a more evenly illuminated picture. They do, though, demand a low level of ambient light. Thus, a matt screen can be the ideal choice for a fixed installation, but you are better advised to choose a more reflective screen for a portable set up. The next chapter examines back projection screens in more detail, as they are often the best choice for multivision shows.

Whatever you choose, be sure that you have a suitable screen available at every venue. It is unwise to rely on white walls – they tend to get painted without notice. In general, the screens supplied for home use are unsatisfactory. Few are large enough, most are rather unstable, and only the most expensive are sturdy and long lasting enough to stand repeated use. Much better, choose a commercial fast-folding screen to project your multivision on to.

Few screens require special treatment. However, they must be handled with the greatest care. Many have easily damaged surfaces, and all are susceptible to dust. Keep your screen packaged in accordance with the maker's instructions when you are not using it.

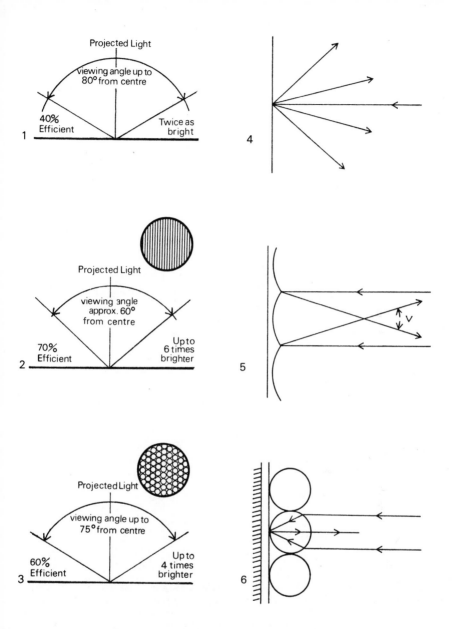

Types of front projection screen surfaces. 1, Matt white. General purpose surface suitable for wide rooms. 2, Lenticular. Projected light is returned in a flat cone of a viewing angle approximately 60° from centre. Fine lines are embossed vertically on a silver mirror surface. Their prismatic shape reflects the light in a concentrated area. Useful where ambient light is high. Six times brighter than an ordinary white wall. 3, Beaded. Fine diameter spherical beads are coated on to a non-fade glue surface to produce a beaded screen. Gives a good brilliance over a wide viewing angle. 4, Matt white. 5, Lenticular. V: Viewing angle. 6, Beaded.

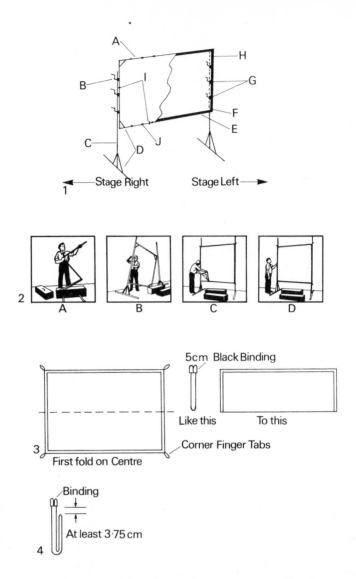

5cm Black Binding

Like this   To this

Corner Finger Tabs

1 ◄—Stage Right   Stage Left—►

2   A   B   C   D

3   First fold on Centre

Binding

At least 3·75 cm

4

Assembly of free-standing fast fold screens. 1, A: Folding frame; B: Crank bolts; C: Folding leg; D: Corner braces; E: Snaps for skirt; F: Black fillet (4 corners for rear projection surfaces only, to mask corner brace); G: On back of frame—blue bolt holders; H: Screen border; I: Hinge and hinge locks; J: Fasteners for screen surface. 2, A: Unfold screen frame and legs. Secure hinge locks and corner braces; B: Insert crank bolts through legs into screw holding device on rear of frame. Screw in until snug. Height may be adjusted by 6 inch increments; C: Snap screen surface on to frame. Note 'top' and 'bottom' markers. D: Raise unit to vertical position. 3, Care of clip-on screen surfaces. Unfold screen surface carefully. If surface is cold, allow to warm up to room temperature first. Clean only with a damp cloth and mild soap. Rear projection surfaces need special care. Do not allow them to come into contact with any painted, varnished or plastic finish. Do not unfold them on the floor where they will be in contact with dust. When packing, fold along long dimension first. The first fold is binding to binding. 4, The second and subsequent folds must *not* overlap screen surface to binding. When sheet is narrow enough to fit into carrying envelope, fold cross-wise, binding to binding.

Image illuminance

When we speak of the brightness of the projected image it is important to avoid confusion between the following three measurements:

Light output from the lamp This can be misleading because it does not take account of the optical system nor of the reflective qualities of the screen in which the image is projected. It is measured in lumens.

Image illuminance. This does take account of the lens used. It is the measurement at the screen but does not take account of the reflective properties of a particular screen, nor the density of the slide.

The equation that relates the object luminance to image illuminance is:

$$E = \frac{\pi t L \cos^4 \theta V}{4N^2 (1 + \frac{m}{p})^2}$$

where E = image illuminance; L = object luminance; t = transmittance; V = vignetting factor; N = relative aperture; p = pupillary factor; m = magnification of projected; and θ = semi-field angle.

The most significant variables in practice are N (aperture) and m (magnification). Either a small aperture or large magnification will produce a relatively dim image.

Projection lenses are made to provide a high transmittance, negligible vignetting and apupillary factor of approximately unity.

Thus the equation can be reduced to:

$$E = \frac{\pi L}{4N^2 (1 + m)^2}$$

Illumination is measured in foot candles, that is, the illumination on one square foot of surface on which there is an evenly distributed luminous flux of one lumen.

Image luminance. This is the measurement of the reflected (or in the case of rear projection the transmitted) image. It is affected by the properties of the screen and by the illumination of the image. The properties of the screen surface will include directional properties, that is, the luminance will vary depending on the direction from which it is viewed. It is therefore measured as being the *luminous intensity* of the surface in a given direction per unit of projected area viewed from that direction. The SI measurement is in candelas per square centimetre (cdm^{-2}). Alternatively, in foot-lamberts.

There are recommended levels of screen luminance which should be used as a guide. This is measured as the 'open gate' luminance, that is, without a photographic transparency present.

In the UK the recommended level is between 30 and 65 cdm^{-2}; average recommended level 40 cdm^{-2}

In the USA the recommended level is between 34 and 69 cdm^{-2}; average recommended level 55 cdm^{-2}

Screen image luminance is not the only important factor to consider. Uniformity of brightness is necessary and the four corners of each screen should be at least 65% of the brightness of the centre.

Subjective tests

Because of the many variables encountered in any projection situation subjective tests are the best guide on each occasion. You should not expect to engage in an exercise of higher mathematics to determine the optimum conditions for staging a multivision show. Instead you can follow a few simple ground rules which will ensure an acceptable picture quality; and follow this up by making adjustments and improvements to the presentation. The following check list may be helpful as an installation guide.

1. Check that the *average* optical axis corresponds to the centre of screen.
2. Screen height should be adequate for viewing by all of the audience.
3. Screen surface should be suitable for shape of auditorium. Wide auditorium needs a low-gain (diffusing) screen. A long narrow theatre requires a high-gain (directional) screen.
4. Check aspect ratio of screen corresponds to aspect ratio of slides.
5. Check that projector throw distance is adequate for keeping keystone distortion to within acceptable limits. With front projection there should be no problem with coping with up to 6 projectors on to a single screen area. With rear projection try to use the longest focal length of lens that space will allow. Registration slides need greater accuracy in line up—use 85 mm or longer.
6. Check that the focal length of the lens is appropriate for screen size and projection throw. Use formula (or lens chart) for determining this.
7. Check that projectors provide adequate light intensity, considering the size of the image required, the length of throw, lens speed and ambient room light. Are high intensity projectors needed instead of standard 250 or 300 watt projectors? Are all lamps accurately aligned inside the projectors?
8. Check that colour temperature is matched, considering the screen tint (if any) and the projector light source. Tungsten lamps give a warmer colour temperature than xenon lamps. Your slides should be colour balanced for the type of lamp used.

9. Check that no member of the audience will be seated closer than twice the individual screen height, nor further away than 8 times the individual screen height.

10. Check that all lenses are of the same focal length and are accurately focused and aligned.

The more rigorously the above rules are observed the better will be the visual quality of your audiovisual presentation.

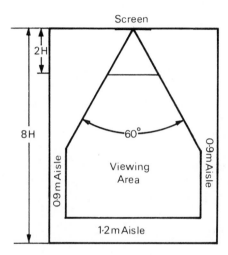

Seating guide. a, 6 sq ft per chair; b, Viewing area 30° each side of centre; c, 1 m (3½ ft) aisle after each 14 rows of chairs; d, No one seated closer than twice height of screen; e, No one seated further away than eight times image height; f, Bottom of screen minimum of 1·2 m (4 ft) from floor.

4   Rear projection screens

Rear projection screens have been used in slide shows since the eighteenth century. It is possibly surprising in spite of two hundred years of experience many current usages of rear projection material do not yield better results.

Why use rear projection?

There are a number of good reasons for choosing a rear projection system.

It provides a clear area in front of the screen where the audience cannot interrupt the light paths of projectors. The projection equipment can be hidden away behind the screen.

Rear projection can provide a brighter and more directional image which makes it useful in higher ambient light situations. On a small display screen rear projection permits close examination of the image. Mirrors can more easily be used to reduce the space needed if the images are rear projected.

The overall size of a multiscreen matrix can be larger. With front projection the projectors have to be mounted well above the heads of the audience so that it is rarely possible to design a vertical floor-to-ceiling screen. Rear projection overcomes this problem. In multivision formats, screen divisions can be more precise and less obtrusive. A rear projection screen automatically becomes one wall of a projection booth, thereby reducing construction costs.

Problems

This is a formidable list of advantages. What then are the drawbacks to rear projection? In fact, the disadvantages occur only if the rules are not observed. Chief among the common faults are these: *Using unsuitable material*, rather than that specifically designed for the job. There are many types of acrylic sheet and gauzes available which, while diffusing light, do not yield the best image definition. They may have some applications as backdrops in theatres for decorative or special effects but they should not be used if the images are an important part of the show. Ground glass is also unsuitable for rear projection.

In particular, no material is suitable for both front and rear projection. Periodically, experiments are made by projecting from both sides of a screen at the same time. They are all doomed to fail because the two screen types require completely opposite qualities.

Overenlarging the image   The size of image that is acceptable depends on the brightness of the projector, the speed of the lens and the degree of darkness in the theatre. If the picture width is doubled, this means that the area that the light has to cover is four times as great. Brightness is reduced by 75% each time the picture width is doubled.

Generally, the picture width in an area with moderate ambient light should be kept to a minimum: 1 m (3 ft) or less. In a completely darkened room, picture sizes of up to $2\frac{1}{2}$ m (8 ft) across can be obtained with a normal 250 watt projector lamp.

Strong light on the screen   To make quite sure that no light falls on the projector side, the projection booth should be completely black, even in the difficult environment of an exhibition. Doors to the booth should be light-trapped and projection equipment should be used that does not require attention during the running of the show.

Incorrect siting   Rear projection works best in long, narrow rooms. Not only is there space in such a room for a rear projection booth, but the audience sit more squarely in front of the screen. Seen at an oblique angle the image can be almost totally invisible because the screen is directing the light towards the front. This is discussed more fully below.

Short projector throw   'I've got a 1 m (3 ft) throw and I want to get a 4 m (12 ft) image. What can I use?' The answer is very simple. Wallpaper. Without a reasonable projection distance the results will be very poor indeed. There are a number of reasons for this.

The widest angle lens which can be used with two projectors is 25 mm. Used in pairs, these lenses will give considerable keystone distortion, due to the distance between each optical axis. With three, four or six projectors covering the same screen area the problem is made more difficult and longer throw lenses *must* be used. A wide-angle lens also distributes the light less evenly across the screen surface, causing a central hot-spot. From the viewing side this will be clearly noticeable.

Wherever possible, choose the standard 85 mm lens for rear projection.

Reversed screen material   The matte side should face the audience. Rear projection material, whether it is the rigid acrylic-based type or flexible material, has a shiny side and a matte side. It is the matte side which yields

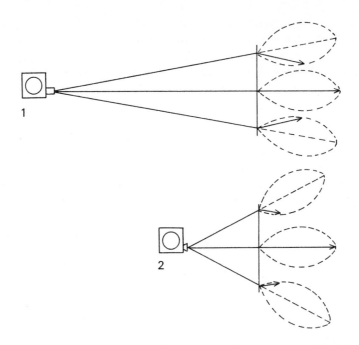

Rear projection screens. When a beam of light hits the rear projection surface it spreads out in a cone-shaped dimension according to the characteristic of the screen. 1, If the apex of the cone is 90° the screen is said to have a gain of 1. This is a low gain and the screen is reasonably non-directional. 2, A medium gain screen, giving a more restricted viewing angle of 60°, but a higher light transmission within that angle, is provided by a screen with a 2·5 gain. 3, A high gain screen limits the distribution of the audience to within a narrow angle of between 15° and 20°. Even within these viewing angles, the brightness of the image will vary.

the image, therefore both sides cannot be matte or there would be a distinct double image. If the matte side faces the audience there is less danger of light from objects in the room being reflected from the screen surface.

Screen flexing In a severe draught, flexible material behaves like the sail of a yacht. This produces an interesting effect of the images flapping in front of the audience but it is usually unintentional. Being both expensive and fragile, flexible screen material is best protected by a large sheet of plate glass, mounted on the projection side. The glass must not touch the screen. If it touches, it produces ugly uneven patches, which no amount of rolling and pressing can disperse. Leave a 62·5mm ($\frac{1}{4}$in) gap between the well-stretched screen and the glass.

The use of plate glass in a permanent situation has the additional advantage of cutting down the noise of the projectors. This can be taken a step further by soundproofing the walls, floor and ceiling of the projection area.

The pitfalls of using rear projection are as numerous as the advantages to be obtained. The screen material which is chosen should be appropriate to conditions under which it will be used. Let us therefore consider the characteristics of the different types of back projection material.

Gain

The gain of a rear projection screen is defined as its transmitted brightness, measured from the audience side, compared with that of a perfect white diffuser illuminated from the same source at the back of the screen. The maximum gain is achieved when the axis of the projection beam is normal to the surface.

When a beam of light hits the rear projection surface it spreads out in a cone-shaped dimension according to the characteristic of the screen. If the

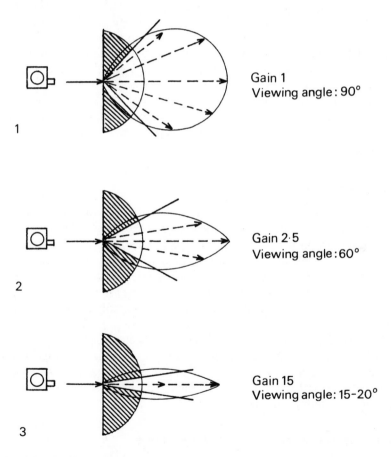

Gain 1
Viewing angle: 90°

1

Gain 2·5
Viewing angle: 60°

2

Gain 15
Viewing angle: 15-20°

3

Screen illumination. 1, Long throw lens. Acceptable image. 2, Wide angle lens. Centre of screen five times brighter than edges.

apex of the cone is 90° the screen is said to have a gain of 1. This is a low gain and the screen is therefore reasonably non-directional.

At the extreme edges of the viewing angle, in the attached examples, the brightness will be 40% less than that seen from the central position. Of course, the projector lens itself will distribute a cone of light to the screen, rather than the theoretical beam. At the edges of the screen the gain will be less and the illuminance diminished. This creates the effect of a hot-spot in the centre of the screen when viewed exactly along the perpendicular axis.

Uneven illumination

Because of the hot-spot effect, the best place to sit in the audience is slightly off-centre, but within the viewing angle dictated by the gain of the screen. With a very low gain screen (less than 1), the spread of light is sufficient to prevent a hot-spot occurring, but the image lacks brightness. The greater the gain, the more intense is the bright central area. This is made worse by using wide-angle lenses. Never use a wide-angle lens with a high gain screen. The designer must bear these criteria in mind not only when considering the horizontal plane but also the vertical plane. If the screen is to be positioned high above the heads of an audience at, say, an exhibition, the gain of the screen should be checked so that it is not too great and it can be viewed easily from the ground. A high positioned screen tends to reduce the hot-spot problem because the perpendicular axis is above the heads of the viewers.

Reflected light

The white surface of a front projection screen picks up any strong light, thereby reducing the contrast and brilliance of the image. This can be cured in back projection by using a tinted screen. Rear projection materials are available in a variety of tints ranging from light grey to black. A screen which is completely black in appearance requires a higher intensity light but it can give a very interesting effect, making the images seem more tangible. For most applications a neutral grey tint is best, enabling normal projector lamps to be used.

Also available are tints which will deliberately change the colour temperature of the light. These should be used with care. A bluish coating will compensate for low colour temperature of an underrun lamp. Mainly, coloured tints, such as the Kodak green Day View type are intended for black and white images in daylight viewing.

Construction of rear projection screens

The normal choice of screen material for permanent installation is a flexible vinyl-based material. This has to be carefully stretched across a wooden frame. Screen material is made in varying widths, the maximum being 3 m (10 ft). Larger than this it has to be joined together. If this is to be done, it is best that the manufacturer or a specialist screen company carries out the work, because the material has to be welded together with heat and even a professional job leaves a noticeable seam. Whenever possible the seam should be made to correspond with the screen separation grid that divides individual screen areas.

Different types of material vary in their degree of elasticity, so you need to test the one you are using to determine how far it needs to be stretched. If it is too loose it will tend to wave about. On the other hand if it is too taut it is difficult to attach to the frame and may even wrinkle at the edges.

There are several ways of attaching the screen to the frame. It may be webbed and eyeletted, that is, given a border of a strong plastic material and fitted with metal eyelets. These are then threaded with cord and the screen is flown with the frame. Alternatively, the webbed border can be fitted with metal press-studs which clip on to their counterparts on the frame. Each of these methods, particularly the second one, is useful if the screen ever has to be removed. For instance, a front projection surface could occasionally be used in place of it.

If the screen is a fixture, it is cheaper to attach the material to the frame with a staple gun. In this case, it needs to be a few inches larger than the frame. Attach one side first, stapling it several times at the centre and then stretching it the required amount on either side. Repeat this for the opposite side of the frame, remembering that it must be stretched in both dimensions.

If plate glass backing is to be added, a 6–12 mm ($\frac{1}{4}$–$\frac{1}{2}$ in) batten should be attached to the frame to provide an air-gap between the glass and the screen. The glass must be on the projection side and the matte surface of the screen towards the audience.

Screen divisions

If the screen is divided into separate areas, without any overlapping projection, a grid should be constructed to obtain precise registration. 'Egg-boxing' will also prevent the small amount of overlap which is inevitable, and displeasing to look at.

Many designs and dimensions have been used for screen divisions. The least successful are those which are too wide and create too large a division between screen areas. If the division is 5 cm (2 in) wide, or even 2·5 cm (1 in), then panoramic effects will look unpleasant—as if they are being viewed from behind prison bars. It is far better to make the divisions extremely small, say 30 mm ($\frac{1}{8}$ in). This does not mean that the grid has to be constructed out of veneer; planed wood, 12·5–18·5 mm ($\frac{1}{2}$–$\frac{3}{4}$ in) thick and about 20 cm (8 in) deep will suffice. In order to get the small divisions, strips of metal about 6·25 cm ($2\frac{1}{2}$ in) deep are sunk into the front of the grid.

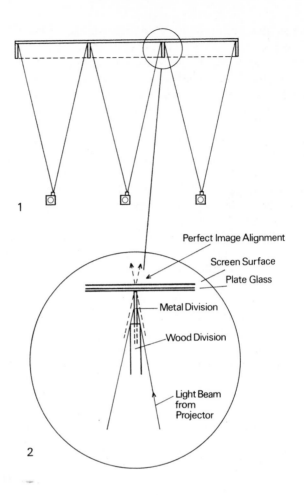

Perfect Image Alignment

Screen Surface

Plate Glass

Metal Division

Wood Division

Light Beam
from
Projector

1

2

Both wood and metal are then painted in matte black and the whole assembly placed against the plate glass until almost touching. If no visible divisions at all are required and yet you still need to stop light from, say, screen 1 spilling on to screen 2, then the assembly can be adjusted to leave a larger gap between the metal dividers and the glass. The edges of adjoining pictures then meet up in a perfect match.

If the screen format is changed for different shows, the grid pattern has to be altered. Central dividers may be made to slot in and out of grooves cut into those which are permanently fixed.

Rear projection cabinets

The greatest difficulty encountered in rear projection is frequently lack of space in the projection area. This is particularly true on exhibition stands

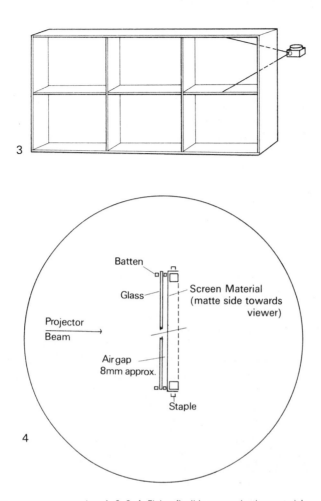

Batten

Glass

Screen Material
(matte side towards
viewer)

Projector
Beam →

Air gap
8mm approx.

Staple

3

4

Rear projection screens: construction; 1, 2, 3, 4, Fixing flexible rear projection material.

where the cost of space can be extremely high. For the smaller installation, an enclosed rear projection cabinet with an integral mirror system can solve this problem.

Mirrors which are suitable for use in projection have a reflective front surface. Ordinary mirrors, silvered on the rear surface, cannot be used because ghost images are produced by the glass. A ray of light striking the glass is partially reflected, while the returning ray is again reflected from the internal side of the surface. The ray is also refracted as it strikes the glass and again as it leaves, owing to the different densities of air and glass. The ghost images will be about $\frac{1}{20}$ as bright as the primary image, enough to be noticeable in graphic slides or in high-contrast images. The degree of

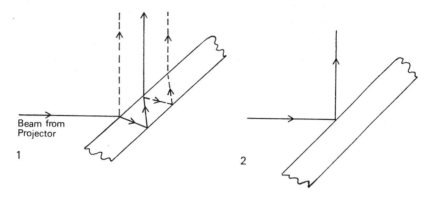

Beam from
Projector

1

2

Mirror surfaces. 1, Rear surface mirror. 2, Front surface mirror.

refraction, and therefore of the image displacement, depends on the thickness of the glass and the angle of the mirror. More oblique angles cause greater image displacement.

A front-surfaced mirror eliminates the ghost images; it also tends to be more fragile, even though the surface is treated with a protective coating to prevent scratches. Producing front-surfaced mirrors is an expensive process and mirrors can account for around 50% of the cost of the complete cabinet.

Alternative materials can be used, such as mirror-foil which has to be stretched over a frame. A good result can be achieved with this although it is more difficult to avoid distorting the optical plane. In practice, mirror foil is also more easily subject to damage. Since any mirror surface will need frequent cleaning, it is best to provide the most stable materials and construction which are available.

Cabinet designs

The quickest way to obtain a near projection cabinet is to purchase a ready-made unit from a manufacturer. These come in many shapes, sizes and colours, and will usually be designed for use with either one or two projectors.

If you are planning an exhibition stand you can get the exact dimensions of the cabinet from the manufacturer and base your design on these. Remember that the front of the screen still needs to be shielded from direct light: A rear projection cabinet bought from a manufacturer will have exactly aligned mirrors, together with a loudspeaker and space for projectors and control equipment. However, if you are unable to find one which is a suitable size and format, it is possible to construct your own.

The design of a complex optical path using mirrors can be extremely difficult if you attempt to do it theoretically. The same goes for the positioning of any projection system. Remember the example of Thomas Edison, perhaps the greatest of inventors. He once asked an assistant, a theoretically orientated European, to measure the volume of a light bulb. When the assistant was knee-deep in mathematical calculations Edison told him to put the bulb into water and measure the displacement. Very practical, and a similar technique can be applied here.

Set up a projector with a slide in it, pointing at right angles to a screen. Use only a moderately short focal length lens, say 60 mm, and adjust the projector throw until you have the exact screen size required. Now measure the projection distance very accurately. With this information you can make a paper template which shows an outline of the projector and of the spread of light from it. The template can be 1:1, that is life size, or 1:2 or 1:4. Show the optical axis, that is, the centre line from the lens to the centre of the screen. In fact, two templates should be made, one showing a side view (elevation) and one showing a top view (plan). In each case label the screen appropriately with indications of left and right, top and bottom.

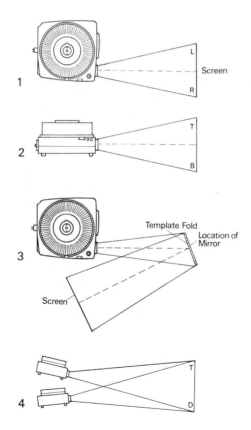

Positioning of mirrors.

Light travels in a straight line. So all you have to do is to fold the light beam at the points where mirrors are to be positioned. Each time you fold the beam the image will be reversed. In this way it is simple to construct a light path with extreme accuracy and minimum calculation.

It is important that the template includes an accurate outline of the projector, with a magazine fitted to it. All projection cabinets are making maximum use of restricted space and there is otherwise a danger of fouling the light beam with part of the projector. Using both templates will ensure that the mirrors can be accurately cut to size and located, in both dimensions. However, do not attempt to angle a mirror in more than one plane. That is, the beam can be bent vertically or horizontally but not in both directions. This will cause a more complicated distortion of the image which would have to be corrected by either tilting the projector or carefully angling another mirror.

Two-projector cabinets

For two projector systems a double template can be made which shows the light beam from both projectors when directed on to a single-screen area. The projectors can be either side by side or one above the other.

The dimensions which are obtained by folding the template are minimum sizes and it is best to increase mirror sizes by 2·5–5 cm (1–2 in) to allow for fixing them in position. Make sure that the extra size of mirror does not encroach on the projection beams.

In practice, the use of more than three mirrors is not recommended because the problems of alignment are increased and there is a noticeable drop in the brightness and sharpness of the image. For each mirror added there is a small drop in brightness, which quickly increases as the mirrors collect dust. The last mirror in the chain should not be too close to the screen because room light can be reflected from it and reduce the contrast of the image.

There are many possibilities for siting projectors, mirrors and screens. Start with the ideal for the particular application and try to fit the projection into the available space. However, it is better to moderate your intentions rather than to sacrifice quality in the projected image.

Construction

A rear projection cabinet must meet certain requirements in construction.

Ventilation must be adequate. Normally no special ducting is required other than the provision of air vents close to the exhaust outlets of the projectors. When positioning the finished cabinet these air vents should not be obstructed.

Easy access to projectors and control equipment is essential. You should,

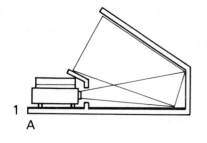
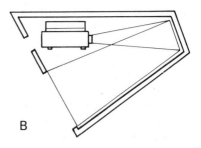

1

A B

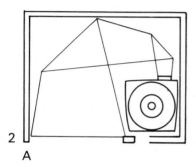
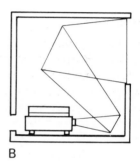

2

A B

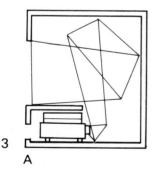
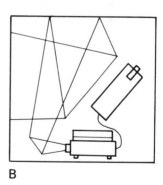

3

A B

Rear projection cabinet designs. In each case care has been taken to retain the projector in a horizontal position. 1, One-mirror systems. A: Open access to projectors. Angled screen for low siting; B: Enclosed projector. For high screen, above head height. 2, Two-mirror systems. A: Front access to projector; B: Rear access to projector. 3, Three-mirror systems. A: Standard focal length lens. Front access; B: Space for control equipment within cabinet.

for instance, be able to remove the magazines without altering the alignment of the projectors. And it should be possible to change lamps with equal ease.

The materials used in construction need to be strong, without being too heavy or bulky. The cabinet should be as soundproof as possible. The interior of the cabinet should be matte black to prevent stray reflections.

Rigid rear projection material is most appropriate for self-contained cabinets. The screen should be recessed to help image brightness.

Loudspeaker(s) should be located conveniently near the screen and firmly secured to avoid vibration to other parts of the cabinet.

Safety is also vital. The whole cabinet should be electrically safe and conform to fire and safety regulations.

If each of these points is observed the cabinet will be an attractive and useful piece of equipment.

Alignment

Continuing the practical approach to making a rear projection cabinet it is best if mirrors are not firmly fastened into place until the alignment has been tested. In most cases it is not necessary to fix the projectors into permanent positions since their weight keeps them aligned unless they are actually knocked out of position. This allows for final adjustments. However, the projectors should be placed in the exact space intended for them while the mirrors are accurately aligned. Line-up slides are essential for this purpose. The mirrors can then be firmly fixed in position with either metal clips or adhesive.

5   Control systems

Used individually, the slide projector is a very humble tool. It can show a series of slides, up to 140 in a tray, giving a brilliant and highly defined image on a front or rear projection screen. In between slides it will plunge the room into darkness, creating a disturbing effect on the audience as their eyes adjust first to the 250 watt illumination and secondly to the 1·5 sec of blackness.

In order to remove the blank screen interval it is necessary to use two slide projectors pointing at the same screen area. Then by the additional sophistication of dimming the lamp of one projector and fading up the lamp of the second one it is possible to obtain a pleasant and natural progression to the display.

At this point one must consider whether it is better that the dimming down and fading up should be independently operated or whether the control of two projector lamps should be linked together in a kind of see-saw arrangement. In practice neither is 'better'—they simply have different applications.

For the simple two-projector set up, it is easier for the dimming down and fading up to be linked together, so that a single command will change the picture on the screen.

Individual projector control is important when more than two projectors are being used, and is discussed in full later.

The dissolve unit

There are many dissolve units available on the market; apart from consider-ations which apply to the purchase of any electronic equipment, reliability and servicing facilities, what should one look for when buying this specialized unit? It depends very much on the particular application that you have, and also possible future applications. It is very easy to spend quite a large amount on a dissolve unit only to find that it does not do everything that you want it to do.

Projector control functions

A dissolve unit can be operated manually or by tape control. It can be used singly or it can be the basis of a multiprojector system in which each number of dissolve units controls a pair of projectors. If you want to build up to this kind of system, check that the dissolve unit can be used on this modular basis.

In addition to dimming down and fading up the projector lamps a dissolve unit gives a command to advance the slide on the darkened projector. This is frequently included as an automatic function, that is, as soon as the lamp has been extinguished the slide change takes place. It is possible, however, to make slide changing independently controllable so that you can go backwards and forwards between two slides, giving an animated effect on the screen.

Alternatively, some dissolve units advance the slide automatically unless a command is given to stop this from happening. This is often known as 'inhibit step'. Some dissolve units give a 'reverse dissolve', that is, they can step the projectors backwards with true dissolves in between. While this may be desirable for a lecturer, it is not desirable for a prerecorded tape programme. Even in a lecture it is possible to get the slides out of sequence using a reverse dissolve.

Lastly, a dissolve unit can give a homing command to each projector which will send them back to zero. This is known as 'reset'. Some units have no reset facility, others have only forward reset and the most sophisticated have forward and reverse reset. Reverse reset is extremely useful in an automatic show where only a few slides are being used. In the homing mode projectors can step only one slide at a time.

Remote control

When used with manual control most dissolve units can operate remotely and a remote hand control may be included in the price. An important point to check here is whether this is a true remote control, capable of operation at, say, 16m (50ft) from the unit. If used as a lecture aid, the dissolve unit needs to be near the projectors at the back of a hall, and the speaker controlling it at the front of the audience. Some remote control systems suffer a drop in voltage that makes the operation unreliable if extension cables are used.

Main features

Each of the above characteristics should be noted when purchasing a unit: Are you always going to use two projectors—or do you want to build up to a 2- or 3-screen show (or larger)?

Do you want the slide advance to be controlled independently?
Do you really need a true reverse dissolve?
Do you need automatic reset?
Do you need shortest-way-home reset?
Do you need a long remote control for manual operation?

Snap changes

An additional question could be added to the list. How sophisticated do you want the visual effects to be? A dissolve in movie-making is usually a luxury because of the expense of complex laboratory processes. In film, however, the movement on the screen takes away the harshness of a 'hard' cut. A good editor will cut on movement so that the change of angle or scene is not unintentionally jarring. With still pictures it is desirable to use dissolves rather than hard cuts because there is no movement to create a flowing sequence. However, there may be occasions when you need to jolt the attention of the audience. The fastest dissolve available is lamp switching and this is known as a 'cut'. The speed of a cut is variable according to the type of lamp and projector used. Even different models of the S-AV 2000 will give slightly different rates of cut, depending upon the lamp inertia, and it is important to use pairs of projectors which are of the same type.

Dissolve units giving a hard cut are available; they operate solenoid operated shutters which can be fitted to the standard Kodak Carousel S-AV 2000 projector. A single command carries out a complex sequence: projector B (for instance), puts the shutter in position preventing light from reaching the screen, the lamp on projector B is faded up, the shutters operate in exact sychronization—opening on B and closing on A. Projector A then fades down and advances its slide. This effect may be a useful addition to the standard dissolves that can be obtained.

Greater dissolve control

Further rates of dissolve are desirable. Commercially available dissolve units fall into two main categories:
1. Continuous tone dissolve units.
2. Impulse dissolve units.

In terms of visual effects, the continuous tone dissolve unit can give an infinitely variable rate of dissolve. This is produced manually by sliding a slider control from position A to position B. You can slide it quickly to give a 'cut' or slowly to give a slow dissolve. Snap changes are never possible with this type of unit. What is happening inside the unit is comparable to a light dimmer. Moving the lever dims down the lamp of one projector and fades up the lamp of the other one. Infinitely variable dissolve units emit a tone which varies in frequency and which when recorded and played back

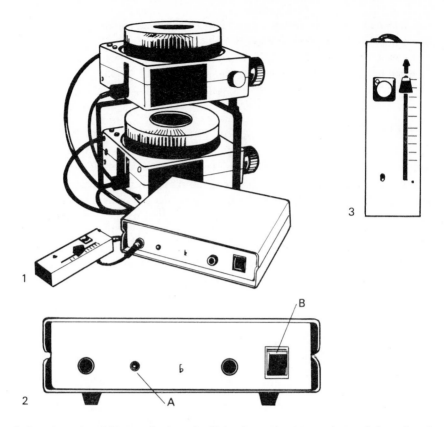

1, A continuously variable tone dissolve unit with hand control and two projectors. 2, Front view of above. A: Signal lamp; B: Mains on/off switch. 3, To use the hand control, slide the slider from A to B, and back again. Microswitches at each end send advance commands to projectors.

through the unit will repeat what you did manually. Slide advance is normally operated by additional microswitches positioned at each end of the slider control. It is therefore possible to go backwards and forwards between two slides without using the slide advance.

The infinitely variable dissolve unit is not suitable for large multivision shows because each one needs a separate control track on the tape. Neither is it easy to make a programme with this type of unit. If you make a mistake with one of the cues then you have to go back to the beginning of the show and start again. It is possible in a studio, with the aid of the correct equipment, to exactly match the frequency and therefore edit the programme by re-recording, but it is time consuming and usually easier to go back to the beginning. Alternatively, you should make sure that whenever you stop the recording in the middle of a show it is always at a point when projector A is at full brightness. When no signal is entering the unit this is the projector that will be on. It is then possible to record the show in sections.

Superimpositions are easily achieved with continuous tone dissolve units. Simply slide the hand control quickly to a half-way position and this shows both slides on the screen at once. With some dissolve units they will each be at half brightness, and therefore, arguably, this is not a true super-imposition. Visually this is quite acceptable because with two projectors you are still obtaining the output of a single projector used normally, if the whole screen area is used in each slide. Other units give a superimposition at 90% brightness. The high brightness superimposition is winning general acceptance.

Reset facilities are available on many continuous tone units. The signal is usually a short break in the tone, for say 5 sec. Turning off the power to the unit has the same effect. It is essential therefore, if you want to pause in the middle of a showing, to turn off the power to the projectors (using the switches at the rear of the projectors) rather than turning off the mains power to the dissolve unit.

The continuous tone or infinitely variable dissolve unit is one of the most widely used, especially in schools and colleges and by the home user. They are frequently less expensive than pulse-type units, and they do not need extra equipment for making a single-screen show.

Pulse dissolve units

The pulse-type of dissolve unit initiates a predetermined rate of dissolve by a single command. This is of great advantage because a predetermined rate has been accurately timed and therefore is not subject to human dexterity in moving a lever. A slide/dissolve presentation will immediately look more professional if there is some order in the types of dissolve that are used. Not only that, but dissolves need to be smooth—and this can only be guaranteed by an automatic pulse-type unit.

A disadvantage of such units is that different pulses are required to initiate different rates of dissolve. Using tape control this will probably mean using complex signals: 'multiplex signals', which are digital pulses containing much information and which rely on expensive equipment for decoding. In practice, simple relays are provided on the dissolve unit itself, these are controlled by a separate piece of equipment, the 'decoder'. However, in manual operation it is simple to operate a number of relays—one button for each. Because tape control may be used in addition, the number of dissolve rates is normally kept to a minimum.

Producers prefer pulse-type dissolve units because it is easier to programme a tape if you are using pulses of short duration rather than one continuous tone. A pulse may be taken out and placed on another part of the tape.

In practice this is done by erasing the incorrect pulse and recording another one.

Remote control for manual operation can be used at a distance from the unit. This remote control will normally be provided with a number of buttons, giving options of the different dissolve rates that are available, together with reverse. The reverse stepping may be a *rehearsal reverse* which will step both projectors back in sequence. Thus you are going back in the show two slides at a time. In order to go back only one slide you have then to press one of the forward buttons. True reverse dissolve is however now finding general acceptance. This reverses the darkened projector by one slide and then initiates a dissolve.

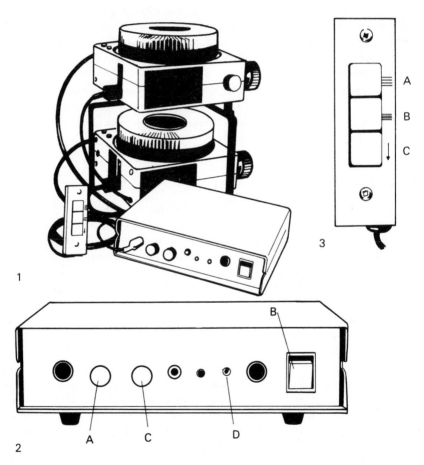

1, Impulse type of dissolve unit. 2, Front view of above. A: Timer interval control; B: Mains on/off switch; C: Dissolve rate control; D: Play/record switch. 3, Hand control for impulse type dissolve unit. A: Dissolve; B: Cut; C: Reverse.

Timing dissolves

Time must be allowed for the darkened projector to step forward after a dissolve. As a guide, some manufacturers state 'dissolve rates' as complete cycle times, that is the time it takes from an input command to when the darkened projector has completed its advance and the unit is ready for its next cue. Even so, to be quite sure that all will be ready, it is best to avoid cue spacing less than half a second longer than the complete cycle time.

The following table shows typical rates of dissolve. The longer the dissolve rate, the longer the reaction time of the eye to perceive it. Thus a 12-second dissolve is already 3 seconds into its cycle before the audience begins to notice it. At just over 10 seconds the dissolve is complete and the projector begins to advance. At 12·5 seconds it is safe for the next cue.

Dissolve cycle time	Time from input to visible effect on screen	Time from input to projector starting to advance	Time from input at which it is safe for next input
2·5	0·3	1·0	3
4	0·8	2·3	4·5
6	1·5	4·4	6·5
9	2·5	7·2	9·5
12	3·0	10·1	12·5

Most dissolve units have a line-up switch. This switches on the projectors. Its prime use is to accurately register the projected images on a screen before a presentation. Line-up slides should be used for this, although, if you are projecting on to a plain white surface without a black border it will be fairly easy to adjust the images until the rectangles (or squares) exactly superimpose. Line-up slides should be stored in position 80 of the slide trays. Access to them is gained by pressing either the individual reverse buttons on projectors or by pressing 'rehearsal reverse' on a remote control. As always, blank slides should be placed at position zero.

Setting up

Setting up a dissolve unit, whatever kind it is, requires a simple methodical approach. The projectors, as mentioned above, are labelled A and B. The convention, which should always be observed, is to have the A projector on the left (as you face the screen) and the B projector on the right. If the projectors are 'twinned' one above the other then the top is A and the bottom B. The first slide in the show starts on A.

Twinning projectors is done when it is desirable to reduce optical distortion to a minimum by getting the lenses of the projectors as close together as possible. The dissolve unit may sit conveniently at the side or it may be on a shelf underneath the projectors. There is no reason why it should not sit on its side, although remember that the projectors must be level and firm.

The dissolve unit will almost certainly need mains power although some will take their power from the projectors themselves. If it takes mains power it in turn feeds mains supply to the projectors and has supply cables attached to it for this purpose. Plug everything in. Place the magazines on projectors making sure that the A magazine is on projector A and B on B. Ensure that there are blank slides already in the gates. Fit lenses to the projectors and switch on. The projector fans are working and one projector has its lamp burning. Go to the line-up slide and focus this projector. Press dissolve, go to the line-up slide on the other projector and focus. Press line-up switch and exactly adjust the positions of projectors until exact superimposition is achieved. This is worth spending a few minutes on; since lazy lining up will ruin the visual effect of the show. Return the line-up switch to its off position.

If there is a reset switch on the dissolve unit press this in order to return to zero. If the projectors start stepping backwards, then your dissolve unit has reverse reset in operation which can be altered, if required, by a switch. If a reset facility is available then the correct projector (A) will automatically light up when this is operated. Always check that projector A is lit before starting the show.

Make sure that dissolve rate switches are in the appropriate positions. On many dissolve units a choice of rates is given although only, say, two can be used in any one show—unless the dissolve rate setting is changed half-way through which is not advisable.

The A projector is shining but the blank slide prevents any light from reaching the screen. An engineering pulse should be given to transfer the light to projector B. This will be included on a prerecorded control tape— if tape is being used. If the dissolve unit is being used manually a dissolve is given to transfer the light to projector B. Projector A advances to position 1—to the first slide of the presentation. You are now ready to start the show.

Integral dissolve units

The circuitry of a dissolve unit may be built in to other pieces of control equipment. For instance, a tape deck and decoder may be included, with the audio amplifier, in a single unit. These package systems, like 'music centres' are matched to a particular specification so that each part of the system is compatible with the rest. If you are purchasing individual dissolve units make sure that the tape deck which you intend to use is suitable for the purpose. There is such a variety of audio equipment available that it is not possible for a manufacturer to ensure that a dissolve unit will work in conjunction with all the different types of tape deck on the market. For this reason there is increasing activity in producing total integrated systems with both replay and record facilities on them.

Autopresent systems

As the electronics which are employed in the control of multivision shows have become more complex so the operation of the equipment by the end user has become more simple. In many situations technical staff may not be available and the user of the equipment may want to be free to do other things besides stopping and starting tape-recorders and adjusting large numbers of projectors and control units.

'Autopresent' is a name given to a completely automatic audiovisual system that will operate in one of two modes:
1. Single shot basis—one run-through of the show, at the end of which the tape stops ready for the next show, slide projectors return to zero, power to projectors is turned off.
2. Continuous running—sequence as for single shot mode with an automatic start for the next showing.

The operation of audiovisual systems in these modes has wide applications in industry and commerce in visitors' centres, showrooms and exhibitions. Necessary to the concept is that the audiovisual system should be permanently installed. Best results are obtained if a designer is used to create the environment in which it is to be seen, if a producer is employed to make the programme and if engineers install the equipment itself.

The scope of autopresentation

Autopresent systems are not limited to small slide/tape shows. For instance, movie projectors may be incorporated, providing they are fitted with a reliable continuous-loop film transport. Houselights are easily connected into the system so that with a single push-button start, the lights will dim at the beginning of the show and fade up automatically at the end. Spotlights may be automatically switched on during a show and curtains may be made to open and close. In fact, anything that can be driven by electrical power and started by closing a relay can be included in the autopresent system.

Equipment

Most types of autopresent show can now be achieved by fitting together standard units purchased from a single manufacturer. It is not necessary to go to an electronics genius and ask him to invent a means of controlling this type of show. No doubt he would be able to do it and would welcome the opportunity because it is most satisfying to watch a complex series of operations initiated by a single command. For the smaller show, say, single screen with two projectors and for small multivisions up to four screens,

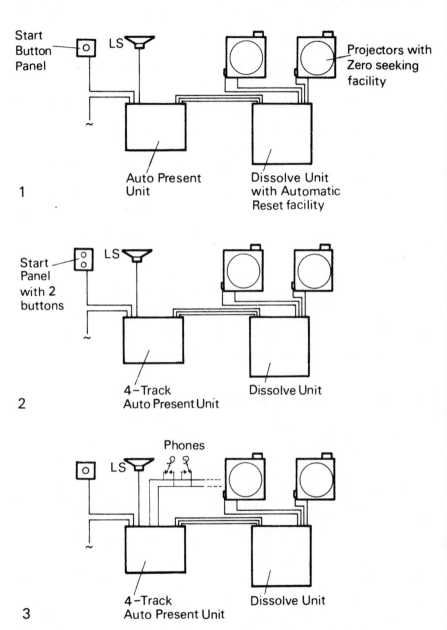

Autopresent systems. 1, Simple autopresent system. 2, Two-programme autopresent system. Programme 2 steps projectors in reverse. 3, Three-language parallel autopresent.

you can buy an autopresent unit that will do all the automatic control.

An autopresent unit contains a continuous-loop cartridge tape deck, an audio amplifier, a decoder and a start-reset circuit. When purchasing this type of unit, make sure that the audio amplifier is of sufficient power for your needs. Most of the package units have only enough amplification for an audience of 20–25 people, because they are used in exhibitions where powerful sound is not necessary.

You also need to add one or more dissolve units, one unit for each pair of projectors. The visual effects that may be obtained are governed by the type of dissolve units. In addition, connect a remote-start-button panel to the autopresent unit and locate it conveniently in your theatre or on your exhibition stand. You require one or more loudspeakers, of the correct impedence, connected to the audio output of the autopresent unit. Finally, but of course essentially, you need projectors to complete the system.

In autopresent systems the best projector to use is the Kodak Carousel S-AV 2000. Fortunately this is widely available especially in Europe where it is manufactured. It has a number of features that are essential to a permanently installed automatic show:

1. Heavy duty construction designed for continuous operation.
2. Circular slide tray with screw-on cover.
3. Built-in reset facility.
4. Thermal cut-out in case of overheating.
5. Wide range of lenses available.
6. Long lamp life when used with dissolve units.
7. Good optical system giving a bright picture.
8. Excellent slide registration.

In the USA the S-AV 2000 is less widely available and the main product in use is the Ektagraphic projector made by Eastman Kodak. The Ektagraphic projector does not have built-in reset facility and this must be added for use in an automatic show. It also does not have a shutter operated 'snap' change and this should be remembered by European producers whose work is to be shown in the USA. However, the Ektagraphic is less expensive (in its US-only form) than the S-AV 2000 and is the most popular projector for multivision shows. It has a faster slide advance mechanism and less lamp inertia than the S-AV 2000. Thus, US producers may find that shows making full use of the projector speed will not work on S-AV 2000 projectors. However, an internationally compatible version of the Ektagraphic was introduced early in 1979.

The dissolve unit provides power to the projectors and other cables connect, as always, into the control sockets of the projectors. Between the autopresent unit and the dissolve unit there is a control wire connection, a mains supply connection, and a reset cable that carries reset commands to the dissolve unit and also reports back information such as 'zero position reached on the projectors'.

Operation

Pressing the start button starts the continuous loop cartridge providing that the projectors are at zero. If they are not at zero, for instance, because the mains supply to the whole exhibition was turned off while they were homing, then the tape will not start until they have homed back to zero. When the tape starts, the sound is played through the amplifier in the autopresent unit. The control signals on another track of the tape are decoded by the decoder in the autopresent unit and these commands are passed to the dissolve unit to carry out. At the end of the show a long pulse stops the tape deck and resets the projectors. When zero has been reached, the power to the projectors and dissolve unit is turned off.

The above operation is in the 'single shot' mode. Continuous run is achieved by giving a permanent start command, for instance, by means of a key-operated switch. The system still goes through the cycle of resetting but then restarts instead of switching off the power.

If, for any reason, the show gets out of synchrony, then it is important for the autopresent system to correct itself without any attention from an operator. This is quite feasible and is a feature of a sophisticated system. Since the tape can stop only at the end of the show this means that subsequent showings will be correct.

It is not advisable to attempt to stop and start an autopresent show. After all, you have installed it because you want complete automation and you must not expect to be able to run a short section of the show over again. If you need this facility, then it is likely that you need a presentation system (as we discuss shortly).

Control tapes

Autopresent units use continuous loop cartridges which are specially loaded with the exact length of tape required. There is always a maximum length which can be used and this restricts the time that a show can run usually to around 18 min. Whereas cartridges are losing the wide popularity they once enjoyed in the audio world, for example in stereo systems, they have a continued use in audiovisual control. Their reliability has been proven and the facility of being able to use an exact length of tape is a great advantage. Continuous loop cassettes have made an appearance in audio systems and will undoubtedly be used for audiovisual control. However, there needs to be more technical development to increase their reliability.

Cartridges use lubricated tape for easy running and the two standards most widely used are, the mini-8 cartridge and NAB standard. The mini-8 cartridge is used with a 4-track format and not 8-track. The track configuration is a feature of the tape deck that is used in the autopresent unit or in the separate tape deck if a large multivision is required.

Using only two tracks of the tape you can obtain mono sound and control. For two-programme operation four tracks are necessary and for this reason the autopresent unit is likely to be more expensive.

Two-programmes

It is a feature of the S-AV 2000 that backward stepping is possible. This may be used to show a second programme with the slides loaded in the trays commencing at slot 80, 79, 78, etc. Thus two shows may be permanently installed in the same projectors and each can be played on demand simply by pressing one of two start buttons. Normally the kind of audiovisual show that is used in the autopresent situation is quite short, therefore there is likely to be spare slide capacity in the trays. It is useful to be able to supplement a main show with another one that may appeal to a particular group of customers or visitors. Show 1 could be about the company's activities in a specific field; Show 2 could explain details about one particular product.

Two-programme autopresent units are available. The normal configuration of the replay head in the tape deck will be Show 1: track 1 audio, track 3 control. Show 2: track 2 audio, track 4 control. Stereo sound is not possible with this system.

Multiple commentaries

Reverting to the single-programme autopresent unit it is possible to give this unit a 4-channel tape-deck for replay of different audio tracks. These may carry different language commentaries. This is a great advantage in many autopresent situations, for instance, in museums and visitors' centres where parties of foreign visitors are taken to see an audiovisual show.

The visual content of the show in each case will be the same but any one of three languages can be switched to the loudspeakers. Alternatively, the language translations can be fed out simultaneously to handsets in the theatre.

Some ingenuity may be required of your producer in matching different language commentaries to the show; one word in German may need several words in Arabic. And these commentaries must lay side by side on the tape with a single control track providing the cueing. If this is too difficult or if the visuals themselves must have a foreign language content then it may be best to use the two-programme system described above, with a separate control track for the foreign language version.

If you need more than three foreign languages then you need more than four channels, capable of being replayed at the same time. Because the standard tape width is 6mm ($\frac{1}{4}$in) it is unlikely that an 8-track autopresent could be made to this standard without the cost price being very high. Currently, the solution is a second tape deck which is started automatically on a cue from the control track of the first deck. This at least enables a translation of the commentary to be given; but for exact cueing a means of synchronizing would be necessary.

Multiscreen autopresent

The same conditions apply to using a multiscreen autopresent system. A single autopresent unit is used to control up to, say, four dissolve units. Beyond this you need separate amplifiers, decoders and tape deck.

The difference lies in the complexity of the encoding on the control track of the tape. For controlling a single dissolve unit you need only a few tone signals to trigger off the different rates of dissolve. When you have to control four dissolve units then a system known as time division multiplex is used. In this system, which is explained fully in the 'Encoding' section, digital pulses are used to provide for more information than a simple tone signal can give. Multiplex signals are measured in terms of size: 8 function, 16 function, 24 function, 56 and 120 function being the most common. Each 'function' is an independent command. The size of the multiplex signal must be large enough to carry the required information for the size of multivision show.

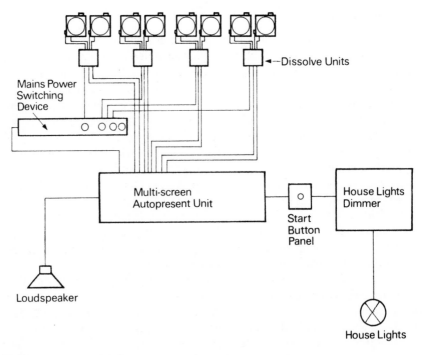

Multiscreen autopresent systems.

An 8-function multiplex signal can give eight simultaneous bits of information. This information is decoded and the dissolve units are told what to do. If 4 dissolve units are being used then each one can receive two commands. More exactly, they can receive one command, or another command, or both together. In this way it is possible, using an 8-function multiplex signal to tell each of four dissolve units to carry out three (not two) separate operations.

So, if your multiscreen autopresent unit has a built-in 8-function decoder you will be able to connect up 4 dissolve units, 8 projectors and be able to get two rates of dissolve and one other effect, say, an instantaneous picture change, or 'snap' change.

Power switching

Small single-screen autopresent units have the capability of switching off the power to projectors at the end of the show. However, a larger system requires another means for this operation, because the switching capacity that is needed is normally too great to be fitted as integral to the unit. Power switching battens are used. These are plugged into the mains supply and distribute power to the dissolve units. A cable connects the power switching batten to the autopresent unit where a low voltage control output of 24 volts, 250 mA operates a remote power relay. In this way cabling, which should always be kept to a minimum, is more conveniently located.

Timing shows

During an exhibition an audiovisual show lasting, say, 10 minutes may be required to run every 30 minutes, unattended. In this case an interval timer is used. This may incorporate a time clock to determine the periods when the show should be active, i.e. during exhibition hours. The interval timer will then be set to start the show during the active period every 30 minutes. This combined unit runs from its own 24-hour mains supply.

Lighting effects

One of the features at an exhibition which is often required is the ability to spot-light a particular product during the course of an audiovisual show. Indeed, many such shows dispense with slide projection altogether and have a programme of recorded sound and a variety of lighting effects.

This may be achieved, again using standard units designed and made by leading audiovisual equipment manufacturers. Light switching may be used in one of the autopresent modes, giving completely automatic operation.

Whatever type of light is used, it may be controlled by a dimmer or directly switched on and off. The most convenient way of controlling this is by means

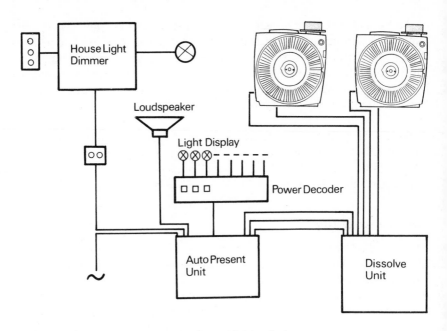

Single-screen autopresent system with programmed lighting display.

of a special power decoder. This enables full mains power to be connected directly to the relay panel supplied with it. Power decoders are modular units, each using 8 multiplex functions. If your projection system is already using functions 1–8, the power decoder can be switched to functions 9–16. A second power decoder could be added to the system and switched to receive functions 17–24. And so on; this is a convenient add-on system that can control not only lights but electrical motors and loudspeaker switching.

A power decoder containing 8 switching relays can decode a multiplex signal and act on the commands contained in the signal. Therefore, it is possible to use it in conjunction with a small autopresent unit to provide a small light and sound show. The two-tone decoder in the autopresent unit would not in this case be used.

Appropriate automatic dimmers connected to the power decoder, can provide dimming during the light display, you need industrial-type dimmers which may have adjustable preset levels. Switching a dimmer on to a preset level uses up a function on the power decoder. Tungsten filament and fluorescent lights require different types of dimmer.

With standard autopresent units houselights control do not require the addition of a power decoder. The dimming down of houselights is accomplished by connecting the houselight dimmer to the start button panel of the audiovisual show. The lights can be set to dim to a preset level instead

90

of right off to provide some ambient lighting. The houselights can be brought up at the end of the show by means of a command given by the tape stop signal. There is therefore no special encoding required to achieve this effect.

It is convenient to have an override control, that is, multiple push button dimmer switches located in one or more places in the theatre. This control may also be remoted to a lectern or special control panel.

The 'Show in Progress' light is one auxiliary display which may be required without going to the expense of add-on power decoders. This may simply be connected to the projector power which, as discussed above, is switched off at the end of the show. However, projector power is not switched off until after all the projectors have returned to zero. If this is too long a period, then another means must be found. The addition of a latching relay unit, consisting of two independently operated relays, will switch the light on at the beginning of the show and off at the end in response to the houselights signal. The relay must have the ability to 'lock on' because it will receive only a pulse signal.

Limits of the standard system

There are three occasions when standard packaged autopresent units may not be the correct system to use, even though you require a totally automatic show.
1.   When the amplification of the audio needs to be much greater, and the overall sound quality improved.
2.   When a large multiscreen is to be installed.
3.   When the format of the show requires individual projector control.

These cases require separate units, including a separate cartridge tape deck, an amplifier and preamps and a decoder. These are available in standard sizes so that they will fit a 47·5cm (19in) rack in which they will stack neatly one above the other. Power switching units are also needed.

Where stereo sound is required, the audio system should be carefully fitted together to give the best sound quality. The tape deck used must have at least a four track playback head and the correct output (usually OdB) for the control signals. Remember that the addition of stereo sound will add considerably to the cost of the programme and is also increasing your equipment costs. Make sure that the location in which it is to be played is suitable for good quality sound reproduction. Museums, art galleries, exhibitions and so forth are not ideal environments for sound reproduction and you may simply be wasting your money. However, in the ideal environment, in a specially built theatre, stereo and even quadraphonic sound is a great boost to the show.

Whereas a 2-channel amplifier can be added to a small autopresent unit that has a 4-track replay head (i.e. the three language version), I do not think this is an adequate substitute for a separate deck/amplifier system. You may be able to get stereo sound and even a great deal of power depending on the size of the amplifier used, but you cannot expect really good quality sound. This can be done only by matching the sound system correctly.

Individual projector control units

Dissolve units, for all their advantages, have one major drawback. If you need to use more than two projectors on a single screen area a dissolve unit can offer no adequate solution. To be limited to two projectors means that the rate of slide advances is also limited, usually to 3 sec between cues.

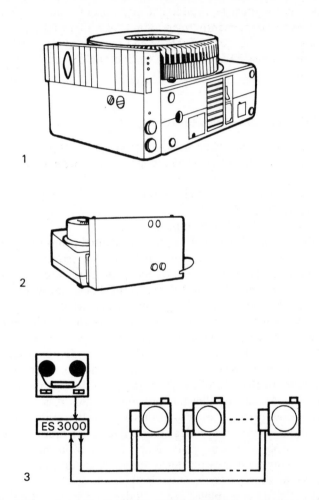

Individual projector control. 1, The Electrosonic 3003 A for Kodak Carousel projectors. 2, The Electrosonic 3003 B for Eastman Kodak projectors. 3, The 'Daisy Chain' linking of the expandable ES 3003 system.

92

Adding another dissolve unit and two more projectors is no real benefit, because each dissolve unit can only tell the projectors to alternate their lamps —not to douse both lights at the same time. What is needed is individual control of every projector in the show, together with a highly accurate facility to fade and dim the projector lamps so that dissolve effects may still be achieved.

There are a number of systems being manufactured which offer this facility. The cost is, at present, marginally greater than using dissolve units but a producer can now use three, four, five or six projectors on each screen area. More than six projectors is difficult because of the distance between the projector lenses in the installation and the consequent distortion of the image on the screen. However, there is rarely any difficulty in using three or four projectors and rapid picture changes are then possible. Not only that, but the increased slide capacity allows for longer shows without magazine changes.

In order to obtain the necessary fading accuracy for individual projector control a device now widely used in industry has been introduced to audiovisual control: the microprocessor. It offers the following advantages:
1. Smallness. The size of control units is reduced dramatically.
2. Economy. The cost of control units offering many more effects can be kept down providing they are manufactured in large quantities.
3. Accuracy. Digital accuracy is now possible with crystal controlled precision.

With the older design of the dissolve unit the rates of dissolve required a number of control channels from a decoding unit in order to carry out its effects on command. More effects meant more channels. With microprocessor control the rates of fade are built into the unit. They are, in effect, memorized by the microprocessor. The actual fading of the lamps is still achieved with thyristor dimmers, but now only three channels are needed to control eight different rates of fade.

Command channels

A	rate 1
B	rate 2
C	rate 4
A + B	rate 3
A + C	rate 5
B + C	rate 6
A + B + C	rate 7
(none)	rate 8

Since the control unit remembers the fade, as soon as it receives its instruction it will carry out that particular effect. This means that the rate codes can be used to give instructions to another projector immediately afterwards.

An example

A particular product which uses microprocessors is the Electrosonic 3003 system.

Each 3003 unit weighing 900 g, clips to the side of an S-AV 2000 projector or to the back of an Ektagraphic projector. A single cable connects them in a daisy-chain fashion, forming a ring of projectors, with both ends of the ring being a demodulator unit. The demodulator, or show controller, receives the multiplex signals from the tape deck, checks them and transmits them in square-wave around the link.

A maximum of 28 projectors can be connected in a ring. For greater flexibility, two rings can be used with a single demodulator giving a maximum of 56 projectors.

The system allows for fade rates of 2·5, 4, 6, 9, 12, 16, 20, and 30 sec together with 'cut' (lamp switching) and 'hard cut' (snap change using shutters on S-AV 2000s). Each rate of fade has a different built-in lamp cooling delay time.

Another effect which is obtainable is 'hold fade'; this allows the projector lamps to be dimmed down to any point and held at that strength while other effects, for instance, superimpositions, take place. After a 'hold fade' the projectors can dim down or fade up at any chosen rate.

The microprocessors also memorize different flashing rates. The slides on each projector remain in the gates and the projector lamps flash according to the rate chosen. In all, there are sixteen different flashing and snap/flashing effects. In each case only a single cue is needed to start the projectors flashing.

1. On/off flashing of all (or any number) of projectors in unison.
2. Alternate flashing (odd and even numbers).
3. Cycle flashing around a group of three projectors, one at a time.
4. Cycle flashing around a group of four projectors, one at a time.

These flashing effects can be at two different speeds:

1. 150 milliseconds per cycle.
2. 300 milliseconds per cycle.

They can also be operated by either:

1. Lamp control
2. Shutter control (with S-AV 2000 projectors).

In total, therefore, there are 16 effects with flashing and snap/flashing, in addition to 9 fade rates, plus snap and hold fade—a total of 27 effects.

Another feature of the system is that information is sent from the projector back to the demodulator. Thus, if a lamp has failed, or if the projector is still homing (resetting), this information is displayed on the show controller.

Each unit has a LED (light emitting diode) display which indicates exactly what the unit is doing at all times. A lamp lights when a command is received from the tape. A digital display shows which particular effect is being carried out, whether that projector is homing or whether it has been told:

'don't step'. The same digital display is used to show which slide number has been reached. A switch may be operated to display the slide number, in which case the lamp is kept dark enabling the projector to keep step while the lamp is changed during a show. When using a whole bank of projectors there is obviously a greater risk of lamp failure and it is important to be able to easily replace these during a show.

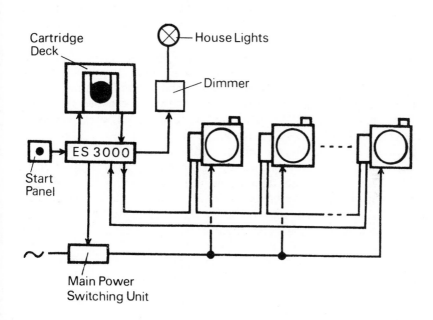

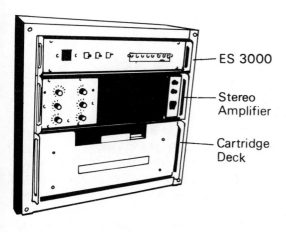

The ES 3003 system in autopresent mode for exhibitions.

Each unit may be set to a number from 1–28 to correspond to the projector in the show which it is controlling. It receives all the information for the whole show but selects that part of the message which is relevant to itself and passes the rest on. The signal is sent in both directions around the single, five-wire cable ring. Therefore, a single bad connection or break in the loop will have no effect on the running of the show.

A line-up switch can be used to switch on individual projectors during setting-up for the display of alignment slides. If these are accidentally left in the 'on' position, the system will ignore the instruction if it is on continuous run in an autopresent situation.

The demodulator (show controller) has a programme start button which gives a choice of programme A or programme B. For programme B it will step the projectors backwards starting with slide 80 and continuing 79, 78 and so on. In either case it will issue an initial step command when a programme start is given. This performs the function 'engineering pulse' in removing the blank slides from the projectors.

A power switching relay is normally used to enable the power to be switched to the projectors. A power switching batten will govern a number of projectors. Thus, when a programme start is given, a command is sent to the power switching unit, the projector fans start up and each projector takes an initial step—forwards for programme A or backwards for programme B. The first signals on the magnetic tape are then counted by the show controller and transmitted in square wave around the link. At the end of the show a 5 sec long burst of set level tone recorded on the tape gives an end-of-show signal to the show controller. This then sends out homing signals until the last projector has homed. One second later the power to the projectors is turned off. The show is now ready for running again.

Inside the system

A microprocessor has four parts:
1. Central Processing Unit (CPU)
2. Read Only Memory (ROM)
3. Random Access Memory (RAM)
4. Input/Output System (I/O System)

The ROM contains the permanent memory store of the unit. This circuit is specially manufactured for the task it has to do and its programme, in the ES 3003 system for instance, has over 2000 instructions.

The CPU uses the ROM as a reference and its function is to 'flick through the pages' of the memory until it finds the relevant set of instructions. For instance, it will receive an input from the projector number switch and will work out which bits of the message are relevant to that projector.

The RAM is a temporary memory which can be altered as well as read. It provides a store for partial results during calculations.

The input/output system tells the processor what is happening and passes on the instructions of the processor to the projector.

There are many inputs for the interface system to sort out:

1. Show commands via the link line from the tape recorder. These are opto-isolated as they enter the unit which acts as a safety device, preventing, for instance, a single pinched wire in any projector from ruining the whole show. The command signals are then interpreted by a serial to parallel converter, a UART (Universal Asynchronous Receiver Transmitter).

2. Synchronizing signals at each zero crossing of the mains supply. Because the dimming is achieved by phase control this is essential information for the computer.

3. Input from homing microswitch on projector that detects the zero position of magazine. This microswitch is already fitted to the S-AV 2000 projector.

4. Input from line-up switch.

5. Inputs from projector number switches: two switches, tens and units.

6. Input from lamp fail detector.

The output interface channels commands to different parts of the unit. The system has the following output wires:

1. Outputs to forward and reverse step controls on the projectors.

2. Output to snap shutter solenoid (if fitted).

3. Output to show controller of information about lamp failure and homing status.

4. Outputs to firing counter for lamp fading.

Digital control of lamp brightness

In the ES 3003, a number of light values cover the range between full projector lamp brightness and total darkness. The scale goes between 0 and 240. An 8-bit firing counter fires the triac at 240 different timings. Each step is so minute that it is not perceived by the eye. This system is also very accurate so that all projector lamps can be stepping up their brightness in unison with each other.

The CPU refers to the permanent memory to find an increment number to add to the light value. This increment number is added to the light value every $\frac{1}{100}$ second so that the firing counter is delayed to give a particular rate of fade. The delay takes place at each half cycle of the mains supply ($\frac{1}{100}$ second on 50 Hz supply). There is also an alternative table in the memory for 60 Hz supply. The firing counter is clock driven by a quartz crystal so that its delay is accurately measured.

There is a non-linear relationship between light value and firing delay for phase control. On old dissolve units the fade up and fade down rates had to be manually adjusted. On newer dissolve units the subjective effects of dimming are taken account of in the manufacture of the unit by the use of diodes.

With computer control all this information is stored in the memory, and the microprocessor finds the correct delay for each light value.

Slide advance

On the ES 3003 system the slide automatically advances when the lamp has been extinguished. Alternatively, a 'no-step' command can be given to retain that slide in the projector gate.

However, a lamp requires a cooling delay time. After a fast fade it needs a longer cooling delay than after a slow fade. On older systems a compromise had to be made because of the complexity of the information required in order to command slide advances at a variety of intervals after the lamp had been extinguished. On the 3003 system this is no problem because complex information is stored in the permanent memory and the lamp cooling delay is different for each rate of fade.

Performance control unit

In large installations a Performance Control Unit displays a complete set of information. It will show which particular projector lamp has failed and will have meters showing the number of hours during which the equipment has operated and the number of shows given. It can also supply a timed automatic start for continuous run shows in public exhibitions.

Houselight dimming

The show controller, or demodulator, has a dimmer socket for the connection of an automatic dimmer. At this outlet a pulse is supplied for dimmer down at the beginning of the show and for dimmer up at the end. This is known as a 'housekeeping function' and it does not use up any programming functions in the system. Therefore no special pulses have to be put on the tape. Because the show controller issues pulses for dimmer control, a manually operated control panel can be wired in parallel for control of houselights when the show is not being run.

Automatic control of movie projectors

Combining the media of slide and movie projection within a single presentation has grown in popularity and many multivision shows now include one or more movie projectors. Technically, the inclusion of movie is as straightforward as controlling a slide projector. Aesthetically, mixing the media is more difficult, and film needs to be carefully integrated into the show.

Control: stopping and starting

Starting a movie projector 'on cue' is usually under multiplex control from the master tape. That is, one channel of the multiplex signal is designated as the 'movie start' channel. When this is in operation a relay is activated which in turn switches on the projector.

Stopping a projector is best done from the film itself, either by a notch and microswitch or by using a cue tone on the magnetic (or optical) track intended for sound.

A latching relay unit may be purchased from a manufacturer of control units for the provision of a convenient relay set. This should be set to the locking position so that the relay will remain 'locked-on' for continued running. The microswitch on the projector unlocks the relay when stopping is required.

It is important that the multiplex start pulses are of sufficient length to ensure that the projector starts. A full 5 sec is usually necessary. If a memory programmer is being used, the start pulses should be repeated on subsequent cues until the 5 seconds have been allowed for. No movie insert should therefore be less than 10 second duration.

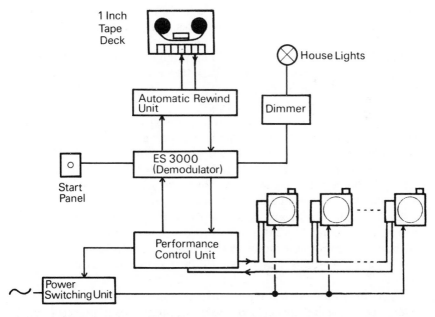

Automatic system for large multivision shows. Additional performance control unit shows information about equipment performance. It monitors each projector and will indicate lamp failures and identify any individual failure on an LED display.

Stopping by means of a roller and microswitch on the projector has the disadvantage that each film print has to be correctly notched. If it is intended that the film will start and stop several times during a show, or if a number of prints are being produced, this can be a time-consuming task. Most 16mm projector manufacturers can supply a roller/microswitch attachment.

The lamp of the movie projector may be independently controlled by means of an automatic dimmer. Again, a small electrical modification to the projector is necessary.

Where there are several stopping points on the film it is more convenient to record a cue tone on the film's soundtrack. A stop cue of 150 Hz and 450 millisecond duration is finding general acceptance. A cue tone generator is used for this purpose. The movie output for the cue tone receiver should be taken from before the volume control.

Control: synchronization

Film inserts often do not need to be exactly synchronized. Providing that they can be started on cue and are reasonably short (2–3 minutes), then the projector speed is sufficiently accurate. Lip-sync may be provided in the normal way, by playing the soundtrack on the film rather than from the master audiovisual tape.

However, there are occasions when more accurate synchronization is required: when more than one cine projector is used, or when slide changes are exactly cued to the film. In those cases, a digital interlock system is required.

Digital interlock is a method of controlling the speed of the motor synchronized by a series of digital pulses. The projector is therefore under complete control from the tape recorder. A package system is made for Elf (Eiki) and Bell and Howell projectors.

The system comprises two units: the stepping motor which is fitted to the projector and an electronic power unit to provide the pulses. The motor requires 200 pulses to achieve one complete rotation of its shaft. The shaft runs at synchronous speeds of 1440 and 1500 rpm and the input driving signal is at 4800 and 5000 Hz. By changing the frequency of the input signal the motor speed can be continuously changed from 1 rpm up to 2800 rpm. Normally the driving pulses are locked to the power supply which ensures accurate running at 24 or 25 fps. A recorded sine wave from a tape recorder may be used to control the motor. Alternatively the control unit will generate its own pulses. Absolute sync is retained in spite of repeated stopping and starting.

Projectors need modifications in order to house the special synchronous motor. The motor can normally be housed within the main projector casing.

Multifilm presentations are possible with the Edit (trademark) system, giving accurate interlock over 2 hours running time. Film editors also find the system useful because it gives synchronization between tape and film without the need to visit a dubbing theatre.

6   Encoding systems

The word 'encoding' implies a degree of complexity in the process of producing signals for control of an audiovisual show; this was not always so. The earliest forms of audiovisual automation used crude and often unreliable means of automatically changing images on the screen.

Basically, to program an audiovisual show means to link the slide changes to a prerecorded soundtrack. The slides and soundtrack are synchronized, and the electronic unit which achieves this is sometimes called a synchronizer. In the days before tape recorders one of the few means of cueing a slide change from a soundtrack on disc was to have a moment's silence in the audio. No music, no speech, just a silent prayer by the producer that the equipment would respond correctly. However, towards the end of the 1940s tape recording was introduced and enabled soundtracks to be more easily made for individual shows. It also allowed more programming methods: metal foil, punched holes, sound pulses on the tape, etc. Anything was used which could initiate a command signal to a projector.

Tape recordings

As tape recording became more sophisticated with the development of stereophonic twin track recorders it became possible for the first time to separate the command signals from the soundtrack itself. Thus the *cue track* or *control track* was born. Apart from an enormous increase in reliability and the recent moves to standardize the sound pulses which are used, the principle of audio pulses on a second cue track has remained. For control of single screen shows there is no better method.

With a single projector, a pulse initiates slide advance. With two projectors and a dissolve unit the pulse triggers a dissolve between the projectors. The whole cycle includes, after a suitable lamp cooling delay period, advancing the slide on the darkened projector. Only a single pulse is needed.

However, producers and users of audiovisual equipment soon required more than simple slide advances for their shows. The market demand in the 1960s grew in two directions: toward increasing use of multiscreen and multiple projectors; and toward a greater demand for many different visual effects: different rates of fade, superimpositions, etc. A simple pulse was therefore inadequate to meet these demands.

Manual control

To begin with, complex shows were staged without automation but with a great deal of human ingenuity and dexterity. Projectionists became adept at controlling great numbers of projectors manually, keeping one eye on the script, one eye on the screen, and the other eye on the equipment. They were also expected to have an equal number of arms if they were to success-fully complete a show. But this type of spectacular, staged at sales conventions and other business meetings required a degree of reliability that had so far been unobtainable. The budgets were there and engineers were invited to cooperate to produce equipment to control the shows.

Manufacturers responded in different ways. The need was for a method of encoding which could command several effects to happen at the same time. One solution was the use of punched tape.

Punched tape programming

In the early 1960s, on the lower slopes of the electronics boom curve, punched tape had a better image than it does today. It was, and still is, a good system.

A punched paper tape system consists of a programmer which punches a row of holes across the tape corresponding to the effects required. The number of effects that could be obtained depended on how many rows the tape could accommodate. Eight rows of holes is normal. The first stage of programming took no account of synchronizing the interval between effects. This was done later by making a second *real time* tape. 'Real time' means the actual playback time of the show as experienced by the viewers.

The reason for breaking the programming up into two stages was to enable the initial encoding to be done at leisure—although 'at leisure', as a phrase, does not seem to have caught on in the hectic world of audio-visual.

The result of punched-tape programming was that many effects were made that could not possibly have been achieved in the real 'real time' of the presentation. It had other advantages too. It is reassuring to the producer if he can actually see the commands. If a hole is not there it is quite obvious. Magnetic tape programming is rather more mysterious. Is the pulse there? Is is the 'right level'? Is there 'tape drop-out'? Is the pulse long enough? But if you can read a punched tape you can see at a glance if something is wrong.

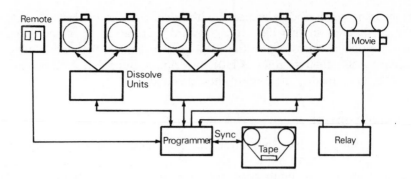

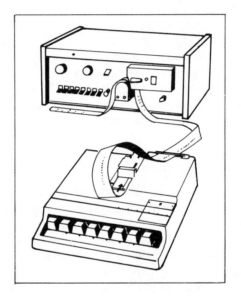

Paper tape programmer.

Eight channels were capable of controlling eight individual projectors or eight single-speed dissolve units. With the appearance of multispeed dissolve units, at the end of the 1960s, a mere 8-command channels were insufficient. If you wanted to use 4-speed dissolve units, then you were immediately reduced to a 2-screen show.

Programming logic was introduced to cope with the demand for more channels. Combinations of holes in the punched tape gave the alternative commands. The first punched tape system to use this was the Media Mix Programmer made by Spindler & Sauppe in 1970. To give an example of how the simple logic works, imagine a set of traffic lights red, amber and green. When the lights show individually they have distinct meanings.

However, a combination of amber and one of the other two (usually red) can convey a fourth message. When all three show together—or none at all—they convey another distinct message—'These lights are not working'.

So the principle was not new, and is certainly older than computers. But it was extremely helpful in solving the problems of multiprojector and multidissolve control.

There was, however, and still is, a major disadvantage. While the physical handling of fewer channels is helpful, the mental handling of them is as complex as before. More so, because you have to learn all the different combinations. Intelligent lay-outs of control panels may reduce this, but an audiovisual programmer (the human kind) needs to have a mind which enjoys the almost masochistic desire to encode: that is to translate simple messages into complex messages.

More recently punched tape programming has become highly sophisticated with the introduction of timing codes that enable a series of effects to happen with an accurately timed interval between them. Rapid slide changes, a fraction of a second apart, produce some startling effects on the screen. These effects are now programmed with electronic keyboard punches. Duplicators are also available which can reproduce punched tapes automatically.

Magnetic tape programming

Alongside the development of punched tape, magnetic tape programming was brought to a peak of ingenuity. Its reliability, at first in question, is now undisputed. For single-screen shows it is now almost universal and as a playback system magnetic tape codes will be around for a very long time.

For the single-screen show there are two main systems in use: continuous tone encoding, and pulse tone encoding.

Continuous tone programming

Continuous tone, which has been briefly discussed above in the section on dissolve units, involves the recording on to tape of a continuous sound that varies in frequency. At one end of the scale projector A will be lit while projector B is dark. As the tone changes in frequency the lamps make a corresponding dissolve until their status is reversed. This is sometimes called a Q-slide system because the method of controlling it is by a slider control, similar to a slider on a sound mixer desk.

Synchronizing by means of a continuous sound has the disadvantage that it is difficult to edit. You really have to go back to the beginning of a programme and start again if you make a mistake. The show is encoded in real time and several rehearsals may be needed. If an attempt is made to edit the tape, then great care must be taken to ensure that the frequency matches at the join. This may be done with an oscilliscope, but then, if you can afford an oscilliscope you can afford a more sophisticated programmer. Editing may be achieved by joining the tape at the points when the slider control is at one end of its run. This can be successfully done with practice. However, editing electronically by recording over an incorrect cue will not work because the tape has to build up to speed. You will then have a few centimetres of tape with no tone on it. Absence of tone is interpreted by certain units as being a signal to reset the projectors.

There are distinct advantages in using a continuous tone for synchronizing a single-screen show. You can obtain an infinitely variable rate of dissolve. Everything that is done manually on the remote control, a fast fade, a slow fade or, in fact, a fade of any length, is recorded by the tape recorder and mimicked by the equipment on playback. To obtain a superimposition, you move the fader control quickly to the half-way position. To advance a slide, microswitches are operated, either on the control or on the unit itself.

The frequency of the tone varies between around 300 Hz and 2060 Hz. When the microswitches are activated the frequency range is extended by a pronounced step.

Because slide advance does not happen automatically after a dissolve you can return to the previous slide or flash the lamps between two slides giving a form of simple animation. With the application of photographic skills there is no end to the number of effects which may be achieved.

The infinitely variable dissolve unit with its slider control is a single unit which does not require a separate encoder – nor, for that matter, a separate decoder. It does all three jobs. Connect it up to two projectors, run a cable from the tone output to a good stereo tape recorder, and you're in business.

The main application for infinitely variable programming is in education: schools, polytechnics, art colleges, etc., and for the skilled amateur. It requires time, but it is normally less expensive than other types of control. Many of these units probably offer the best value for money on the market.

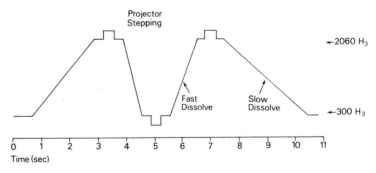

Continuous tone encoding. Diagram shows what happens when slider control of a continuous tone dissolve unit/encoder is operated. Slider moves from 'A' position to 'B' position.

106

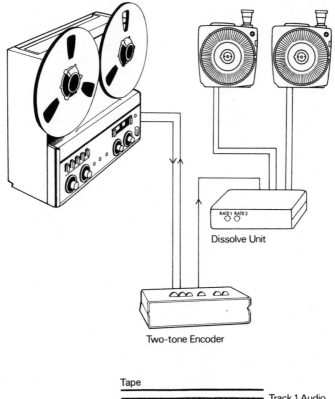

Dissolve Unit

Two-tone Encoder

Tape

Track 1 Audio

Track 2

Track 2 silent except at Pulse

Simple two-tone encoding on magnetic tape.

Pulse tone programming

Simple pulse tone encoding provides the great benefit to the professional producer of easy editing. You can erase a pulse and drop it in on another part of the tape. It offers a very limited range of dissolve effects because a different frequency, or a different length, of tone is required for each rate. These pulses have now to a large extent been standardized as this extract from Publication 574–10 of the International Electrotechnical Commission will testify:

'*Audio Visual, Video and Television Equipment and Systems*'
Part 10: Audio cassette systems

6.   Preferred characteristics and applications of the cue tones.

6.1   Preferred characteristics of the cue tones.

A cue tone shall consist of a burst of a sinusoidal tone. The frequency of the tone is to be preferred to the duration of the tone as the distinguishing factor between different functions. 150 Hz and 1000 Hz are the primary frequencies (see sub clauses 6.2.1. and 6.2.2) with 400 Hz and 2300 Hz as the secondary frequencies (see sub-clause 6.2.3.).

6.1.1. Tolerance

The tolerance of the frequency of the cue tones shall be $\pm 6\%$ when reproduced at rated speed, while the equipment shall be able to respond to cue tones with a tolerance of $\pm 10\%$. The tolerance on the cue tone burst duration shall be $\pm 16\%$ when reproduced at rated speed, while the equipment shall be able to respond to a cue tone burst duration with a tolerance of $\pm 20\%$.

6.2   Applications of the cue tones.

6.2.1   Picture advance and automatic switch-off.

Frequency 1000 Hz.

The duration of the cue tone burst shall be:

–450 ms for visual frame advance only, still picture (one frame)

–100 ms for film strip advance only, animation (one frame),

– 24 ms for motion picture film advance only (one frame),

– 2 s for advance and automatic switch-off for slides and motion picture film.

6.2.2   Pause (temporary programme stop)

Frequency 150 Hz

The cue tone burst duration shall be 450 ms for slides and films.

6.2.3   Other cue tone applications.

Other cue tone applications such as:

random access, programmed instruction, superimposed cue system for advance and programme stop, are under consideration.'

So, at long last, a standard has been fixed. This will be of immeasurable benefit in enabling programmes to be played on equipment made by different manufacturers. Slide/tape competitions will no longer require a dozen different systems for showing the programmes that have been entered. Track formats have similarly been standardized.

It is worth noting that the tolerances for frequencies of cue tones have been set to quite accurate limits. In the early days of tape recording such accuracy was not possible, and quite often cue tones could be anything from 500 to 2,000 Hz. The equipment 'reading' the pulse had to respond to a wide range of frequencies. If the voltage dropped, the tape ran slower and the frequency of the tone fell. Modern equipment is not subject to such variations and the reliability of this particular aspect of multivision is high.

Digital encoding

Multiscreen and multidissolve systems, as we have seen, require many different channels in order to give the required number of commands. Punched tape offered one solution and, with the addition of simple logic, was very effective and reliable. However, it was inevitable that the users of multivision equipment would eventually come to favour a system that reduced all the programming instructions to a single cue track on tape. This was achieved by the development of digital pulse encoding.

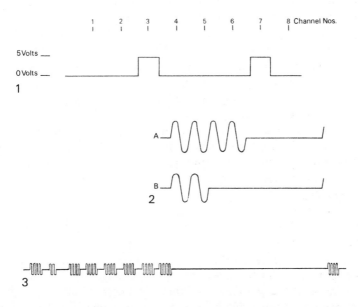

Digital multiplex control signals. 1, Square wave signal. The square wave signal is converted to the division multiplex. This is composed of a train of tone bursts of a 5120 Hz sine wave. Each train has a number of bursts that correspond to the number of functions being transmitted. 2, A: The bursts are of 4 cycles on and 4 off for each function wanted to be *not* active (clear or space); B: And are of 2 cycles on and 6 cycles off for each function wanted to be active (set or mark). Each burst is keyed on or off at its zero crossing point. 3, Each train of bursts is preceded by an inter-train silent gap. For normal speed signals this gap is 12·5 milliseconds minimum. It is usually longer and is chosen to make the entire train repeat at a convenient periodicity
    messages up to 24 channels at every 50 milliseconds
    messages up to 56 channels at every 100 milliseconds
    messages up to 88 channels at every 150 milliseconds
    messages up to 120 channels at every 200 milliseconds
for normal speed.
Where a high quality tape system is being employed the multiplex system can be operated at double speed. The basic frequency becomes 10240 Hz and 120 function messages can be sent every 100 milliseconds.
For use on very poor media such as 8 mm magnetic stripe and to make checking recordings easier, the decoders (and encoders) can be operated at half speed.
The use of a 5120 Hz basic waveform ensures that normal speed multiplex can be used with security with any tape recorder capable of good quality recording. A bandwidth of 1–3 dB from 400–6500 Hz is required.

A digital pulse is not a collection of frequencies, as might at first be supposed. It is in fact a single tone which is given a pattern of on/off signals. This pattern of on/off signals contains many bits of information which can later be decoded, either by the programmer itself or by a separate decoder.

Because the pulse alternates between tone (on) and silence (off) it is not greatly affected by variations in tape speed. In fact, it is common to have a speed tolerance of plus or minus 40%. Likewise, the system of digital encoding —sometimes called 'time division multiplex'—is not dependent upon amplitude or small variations in level. It *is* subject to tape drop-out, and for this reason digital pulses are normally repeated three or four times. The decoder checks the signals, counting the tone pulses in each phrase and comparing adjacent phrases. The combinations of digital pulses and double checking procedures have created a highly reliable control system.

Synchronization with digital encoding requires that the pulses are placed exactly in their correct positions on the tape. This is not difficult with a simple show, you just have to press a few buttons at the appropriate moment. But with a show using many projectors and many different rates of fade, the problem for real-time encoding is magnified to the level of impossibility. For this reason, the larger encoders do not work on a real-time basis. Instead, the combination of bits in the multiplex phrase is arranged on a control panel, each button lighting up as it is pressed, and the whole combination is put on to tape.

Let us take, as an example, 8-function (or 8-bit) multiplex. The signals for each channel are represented by a change in voltage from (in some systems) 0 to 5 volts.

When a channel is engaged the voltage steps up to the maximum level. In graphic terms this can be visualized as in the diagrams. The high voltage steps correspond to the pressing of function number buttons. In practice, the combination is punched out on the control panel and the 'send' button is then pressed in order to put the multiplex signal on the tape.

There are, however, two further stages between setting the combination and the recording of the signal. First, the code goes through a counting process which puts the function numbers into series form. In setting the original combination the operator is putting in parallel information. This is received by the 'shift register' which changes it to a series or linear form. Secondly, because discrete levels cannot be recorded, the square waves have to be converted to a carrier frequency wave form.

The carrier frequency which is used in some systems is $5 \cdot 120\,\text{KHz}$. Each high in the square wave signal is allocated two cycles of this frequency. Each low is allocated four cycles.

For the multiplex signal which goes onto tape, a division of a second is allowed for the complete signal to be expressed. A short 8-function signal will use only a part of the $\frac{1}{10}$ second slot allocated to it. On the other hand, a 56 function signal will take up most of the time slot. Larger signals, carrying even more information, require more than $\frac{1}{10}$ second and the time division is increased to $\frac{1}{5}$ second. A gap is necessary between 'sends' and each new signal commences at the beginning of the time slot.

Because the 120-function signal, which is in effect 15 8-function signals, requires a longer time slot, it is sent at twice its normal speed. That is, the carrier wave frequency is increased to 10·240 KHz.

Decoding the signal requires the multiplex control on the tape to be re-played through a decoder which carries out the reverse process to encoding. The decoder checks the *size* of the multiplex to see whether it is 8, 16, 24, (etc.) function size. It also compares adjacent sends for accuracy before acting on the instructions. The serial information is then converted into parallel information and commands are sent simultaneously to all the dissolve units which carry out fades, dissolves, slide changes, etc.

Multiplex signals are only transmitted when changes are required. In playing back the control track through an amplifier, and listening to it, you will hear the code whenever a new instruction is being given. This makes it easy to edit a tape during silent periods.

Memory encoding

Memory encoders are not indispensable to audiovisual production since the visual effects can be achieved by other means. However, they can be time-saving, enabling the person encoding the show to achieve complex effects very quickly. They also provide extreme accuracy of cueing and they enable almost instantaneous editing of any part of the show.

Memory encoders are intended to programme the big show. The smallest can control nine projectors. Because sound still needs to be recorded on tape and synchronization depends on command signals on another track of the tape there is little advantage in using a powerful computer memory to program a two- or three-projector show. You still have to synchronize it.

Synchronization with memory programmers can be done in one of two different ways depending on the capability of the equipment.

Either, the memory contents of the programmer are activated on a cue-by-cue basis by means of a simple tone pulse on the tape.

Or, The programmer with a built-in clock timer automatically associates each cue with a particular real time code and places a complex signal on to tape.

Each of these methods has inherent advantages and disadvantages. Basically, the first enables a producer to alternate with ease between 'canned' (i.e. taped) programmes and live, speaker support sequences. Each set of slide changes has a cue number associated with it and so you know exactly where you are in the show. Method two provides for greater accuracy in cueing and is particularly suited to programming large fixed-installation shows. Both types of programmer can store the program in computer language on tape. They can both be refilled from this tape store.

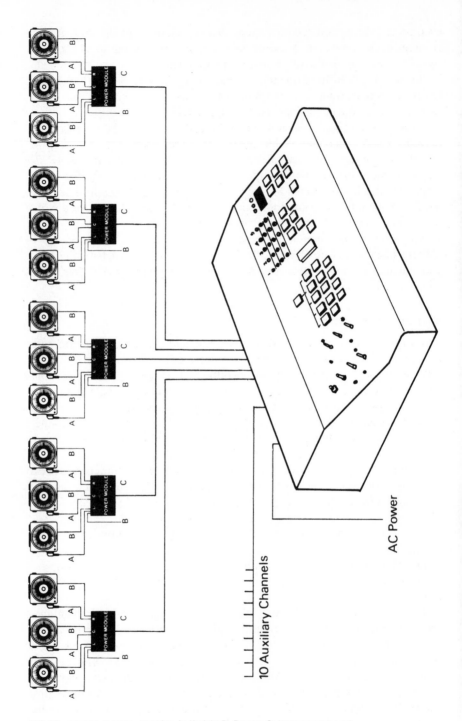

AVL 15-projector memory encoder. A: Kodak; B: Power; C: Interconnect cord.

112

The great advantage of memory encoders is that they work out all the tedious mathematics that is involved in producing a complicated code. It is easier, for instance, to press a button marked '8 seconds', for an 8-second fade, than it is to look up a table which tells you the correct combination of buttons that is required for sending that command to a particular projector. With a simple effect the time saving is not significant but for a sequence involving a succession of rapidly flashing projector lamps then memory encoding comes into its own.

With computer control doing much of the work for you it is possible to indulge in a few luxuries. For instance, it is easier to program a show with projectors connected so that you can see the effects immediately without having to manually step them backwards or rewind a tape. At the touch of a button the projectors seek the correct position and await replay of a sequence or carry out a new, edited, sequence which is inserted into the programmer memory. This kind of 'instant art' has great appeal to anyone producing multivision shows.

There are, of course, dangers in the instant approach. In an audiovisual show it is always the cerebral content which matters: the ideas, organization, and imagination which have been put into it. The skills of the photographers, artists, writers and sound engineers are even more important than those of the person encoding the show. In fact, their ideas should leave the programmer with relatively little scope for ingenious gimmickry. Complex sequences need to be designed before they are programmed.

Each memory encoder has its own characteristics, and production methods need to be adapted to the type of equipment which will be used for the programming. For instance, cue sheets should be designed to cope with the procedure used in addressing a memory programmer. This will make the task of encoding a show both simpler and more accurate.

I have no doubt that memory encoders will continue to be developed even beyond their present high standard of sophistication. These developments should make them easier to operate, because of their ability to process more information.

'Real' computer encoding

Instead of using a computer to make the memory for a special audiovisual programmer you can use an ordinary 'domestic' computer for encoding the show. One system is based on a word processing machine, consisting of an alphanumeric keyboard, video display unit, minicomputer and floppy disc storage. This system uses a new computer language called Procall which is specially designed for audiovisual programming and display.

Procall stands for Programmable Computerized Audio-Visual Language Library, and is designed to communicate with the operator in English. Its conception is a logical step in the development of programming systems because a low-level language computer can easily be updated with new software. In practice, a 'real' computer can cope with literally hundreds of different visual effects, far beyond that of even the most sophisticated memory programmer. With expandable memory it is possible to handle 30 000 or 60 000 cues, enough to contain a whole-day multivision with plenty of room to spare.

The system has a number of processors (or brains) and is therefore capable of carrying out several instructions at the same time. The key to the system is the special Procall program which is made available in disc form and which is loaded in a matter of seconds. Without Procall the equipment is simply a word processing machine. This in itself is useful to a producer because he is able to use it for producing amended scripts, again within a few seconds. The equipment plugs into a variety of standard printers and will give a hard-copy print-out of the entire show. Alternatively, the system can be used as an accounting machine and will even produce mailing lists. If these other tasks are required all you have to do is to change the disc.

As an audiovisual computer the system allows you to type in coded instructions. The code, for example 2D, meaning 2-second dissolve, appears on the video display, together with a full description, i.e. *2-second dissolve*. The VDU shows not only the cue at which you are standing but nine other cues as well. These may be, during editing, five cues forward and five cues back. Previously, other programmers have only shown a preview of the next cue and a review of the last cue. Also on the display is the projector layout and status.

Real computer systems offer a flexibility which enables a producer to design any format of multivision show. He is not tied down to using a standard format of, say, three projectors on each of five screens. Other memory programmers can cope with a degree of flexibility and it is true to say that the more versatile the system the more complex is the programming. More than ever before, special skills are required for encoding audiovisual shows. People with a knowledge of computer language and operation are needed not only in the design of programmers but in actual program production.

Types of encoding systems in current use

1. Punched tape programming
Many different tape widths used. Average is 8-hole tape.
Advantages:
1. Reliable.
2. Easy to read tape without decoding it.
3. Simple to edit.
Disadvantages:
1. Information is not stored in compact form.

114

2. Original tape does not determine real time interval between cues.
3. Limited number of channels.
4. Tape subject to damage.

2. Magnetic tape: Infinitely variable tone
The basic two-projector encoding system.
Advantages:
1. Infinite number of dissolve effects.
2. Simple half-brilliance superimpositions.
3. Easy to achieve 'twinkle' or flashing effect.
4. Dissolve unit itself will encode and decode. Therefore no extra equipment needed.
Disadvantages:
1. Can be time-consuming, because of need to encode whole show correctly in one 'take'.
2. Dissolve rates 'mimick' manual operation of slider control. Can therefore appear ragged.
3. Can only be used for two projectors.

3. Magnetic tape: Tone pulse
The preferred system for two projector shows. Dependent upon frequency and duration of one (or more) pulses.
Advantages:
1. Easy to synchronize with picture.
2. Easy to edit.
3. Extremely reliable.
Disadvantages:
1. Can only be used for two projectors, although some dissolve units may also be used on a modular basis for multiscreen shows.

4. Magnetic tape: Real time digital encoders
For the small multivision show. Maximum normally 16 channels.
Advantages:
1. Simple to use (but dexterity needed!)
2. Synchronizing achieved in one operation.
3. Can be used to alternate between 'speaker support' slides and recorded tape programmes.
Disadvantages:
1. Complex cues require pressing of several buttons at once. This requires practice and skill.
2. Limit to size of show that can be programmed (as a consequence of above).

5. Magnetic tape: Incremental digital encoders
For any size of multivision show.
Advantages:
1. Extremely versatile.

2. Highly reliable system.
3. May be used with dissolve units or individual projector control units.
4. Simple to cue lights, motors, special effects, etc.
Disadvantages:
1. Complex effects can be time-consuming to achieve.
2. Necessity of translating effects into appropriate function numbers before encoding.

6. Memory encoding: Clock/memory
For any size of multivision show.
Advantages:
1. Real time of show is entered into memory and any cue is instantly retrievable.
2. Instant access for editing.
3. Effects are entered into memory—more easily understandable than a function number code.
4. Well-guarded keyboard.
5. Split-second timing possible giving more fluid and accurate cueing.
Disadvantages:
1. Needs to be operated by someone with computer programming skill.

7. Memory encoding: Cue/memory
For medium-size multivision shows. Maximum 15 or 24 projectors.
Advantages:
1. High speed cueing.
2. Easy to alternate between speaker support slides and 'canned' modules.
3. Fast editing.
4. Screen status display.
Disadvantages:
1. Show needs to be synchronized after cues have been entered.

7. Real computer encoding
The use of a word processing machine with a specially written memory for audio visual encoding.
Advantages:
1. Extremely flexible.
2. Excellent status display.
3. Can be put to other uses.
Disadvantages:
1. Expensive.
2. Complex to use.

Cueing the programme

Studio encoding time, tends to be expensive. High-cost equipment and highly skilled engineers are needed to encode the more complex multiscreen shows.

116

However, this expense represents only a fraction of the total cost of a programme because encoding takes only a few hours in comparison with the days, weeks and (sometimes) even months which may be spent in scripting, photography, graphics, etc.

Encoding should be, as far as possible, a mechanical task. The most creative programming engineer cannot turn a bad set of slides into an interesting and purposeful audiovisual show. Encoding is not the audiovisual equivalent of film editing. The equivalent is in the slide sorting stage when a final choice of visual material is made. Encoding is rather the mechanical translation of predetermined cues into a series of command signals.

Therefore, it is the cueing of the programme which is of primary importance. It is the preparation of cue-sheets which, like an architect's drawing, determines the final appearance of the show. First, the show is visualized, then it is cued, then it is encoded.

There has been a tendency in recent times, with the introduction of sophisticated computer memory encoders, to muddle these three operations into a single activity. This is self-defeating, because the time saved by the use of memory programmers is wasted if the person programming the show is constantly readjusting cues in order to pull a show together at the last minute. Audiovisual shows need thought in synchronizing of soundtrack and pictures. They need thought in visualizing of picture sequences. Encoding is a physical, mechanical operation.

Having said that, it is clear that minor adjustments have to be made when encoding and also that it can be constructive to experiment and 'play around' with some sequences in order to achieve clever visual effects. In practice, experimentation with encoding has a high failure rate. For a start, the soundtrack dictates certain parameters beyond which your clever efforts will appear as mere gimmickry. Secondly, images need to appear to be organized. If there are, say, a hundred possible combinations of images within a particular sequence then it is extremely likely that the sequence has not been predetermined, so that *any* image combination/succession will appear to be totally random.

Cue sheets

Whether your show is a simple two-projector presentation or a multiscreen epic it will pay good dividends to draw up a precise cue sheet. This sheet contains all the information that the programmer requires. It has the time, in seconds, running vertically down the graph paper. It also has a column for each projector, or a group of columns for each screen area. The full width of a single column (that is, conveniently, five squares of standard graph paper) represents the maximum brilliance of a projector lamp. The fading up and fading down times of each projector are shown by tapering each block to a point. Note that a gap is left between each block which represents the period when the projector is advancing to the next slide. If there is no gap, then, if the cue sheet is correct, the same slide is to be shown again. Cue numbers, corresponding to slide changes are written in at each new fade-up point. For complete clarity the fade rate may also be written in. On the diagram there are four fade rates shown. Any other instructions, such as 'engineering pulse'—an initial step to advance to the first slide (which is achieved on some systems automatically) should also be shown.

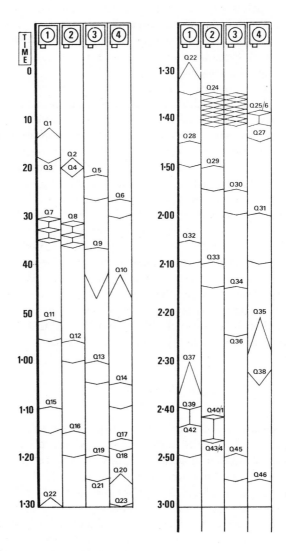

Graphic representation of programming for few projectors.

However, very little other information is required and the cue sheet is therefore kept neat and uncluttered.

The cue sheet contains in graphic form all the data which is needed: that is, the time in seconds, the fade rates, the projector status, the cue numbers, and the minimum stepping time.

You can tell at a glance whether the effects you have chosen are achievable —without the burden of using expensive encoding time for juggling all the components of the show.

If the encoding is to be made directly on to magnetic tape it will be helpful to put a 'clock track' on to one track of the tape. A clock track is a series of discrete pulses which are interpreted by a multiplex lock that displays the programme time on a digital read-out. Each 'tick' of the clock displays time to the nearest $\frac{1}{2}$ or $\frac{1}{10}$ second and displays the correct time even if you were to start in the middle of the tape.

Having a clock track enables you to find individual cues with great ease. Its great advantage is that it *locates the cue in time*. Systems which display only cue numbers have the disadvantage that you have to return to a previous production stage—the script or storyboard—in order to locate a particular cue.

The method described above is the most accurate and ultimately the most professional way of preparing to encode an audiovisual show. Other methods are used, usually for speed of production, which use simplified styles of cue sheets. However, any attempt to cut down on the paper-work means that you have to leave out some of the vital information which is needed for efficient encoding.

7 Scriptwriting

Since a multivision show is the product of many people, each with different skills, the scriptwriter is only one person among many. However, his is the first, and often the least mechanical stage. His job is simply to write a description of the final product. A good script is really a work of conceptual art which, with the addition of the reader's visual imagination, is in itself complete. It is not merely an outline for a director to turn into whatever he wishes it to be. On the contrary, the script should describe precisely the effects that the photographers and sound technicians should try to achieve.

It is important that one person should have some degree of control over all the elements in the show. The scriptwriter is in the best position to fulfil this role. In practice, he is often also the director which is obviously a convenient solution to any conflict of ideas; if he is not, then the director must accept that his role is an interpretive one.

The most important quality of a scriptwriter is his range of ideas. It is quite useless if he writes a commentary full of facts, however accurately and eloquently expressed, if he cannot transmute the subject matter into a new medium. If ideas, integral and sympathetic to the medium, are not there at the script stage, the director may provide some. This may produce a good programme, but it is the director's job to direct and organize, rather than to conceptualize the show. Also, the sponsor will be more than a little surprised to see that the final show bears no relation to the script that he read a few weeks previously.

Diversity is essential

The scriptwriter should also have an immense intellectual curiosity which will sustain him through many different subjects. If he is incapable of being interested in road building or the history of Islam or American furniture, then he will not interpret the subject in an exciting way for an audience. He will not, preferably, have a degree in road engineering or Islamic history or whatever specialist subject he is scripting. If he does, he is almost certain to assume too great a knowledge of the subject by the audience. Multivision programmes are rarely used to disseminate knowledge among elite bands of professionals.

The scriptwriter must, though, be an audio visual specialist with a thorough knowledge of 'the state of the art'. He must know the technology of the medium; what is achievable in the sound studio, the graphics studio and by the camera. On top of that, he must be able to understand the sponsor and communicate with the audience.

Working within a team is an essential part of the scriptwriter's life, but he must stand slightly apart, simply in order to grasp the form and content of the whole programme. In practice, the writer must be involved in the making of the show so that his ideas can be realized. But while he is writing he must enjoy a degree of privacy and quiet so that not only the creative functions can operate but also the equally essential critical functions which will correct and balance the script and reject non-workable ideas.

The script

The script is in words and must be, as near as possible, complete in itself. It is meant to be read and understood by the rest of the production team and by the sponsor. But it is not the show itself. The show will be seen and heard simultaneously, not read from a printed page. Therefore, scriptwriting is not a literary activity, it is an audiovisual activity which is described in words. The whole of this book is an exploration of audiovisual activity, but here it will be useful to examine the particular elements which concern the writer. A professional writer will be able to assume the craftsmanship of the medium in which he is currently engaged. In writing a play, for instance, he knows the limitations of the theatre and writes within them. His work is essentially dramatic and to this end he creates dramatic situations on stage during which the chief vehicles of communication are speech, gesture and limited movement. Battle scenes and big events will take place off-stage and the play concentrates on people, on their reactions to events, and so on.

The novelist is less confined in as much as he can put his drama in any setting providing his powers of description are adequate. And the film writer, too, has immense resources at his command, providing that the dramatic activity and the setting can be adequately evoked by actors and camera.

It would appear that the multivision writer has immediately to sacrifice the most precious element: drama itself. Or does he? Without drama it would be impossible to create an interesting audiovisual show. Drama, or conflict, is the essence of any communicative medium, whether it be speech, strip cartoon, film, TV, factual journalism, multivision, poetry, theatre or music. Music is an excellent example, being pure drama in sound: the development of themes, contrasting rhythms, constant surprises.

Dramatic tension

If you are about to write an audiovisual presentation to promote, say, weed-killer and are wondering what all the highbrow arts have to do with it, then think carefully about the nature of drama and its possibilities. Even a soap powder commercial on television contains a certain element of drama, despite the fact that you never actually *expect* Mrs. Jones to choose the pile of clothes that have been washed in Brand X. Scientific research is full of drama. False trails are followed, there is disappointment, failure, accidents, tragedy. Not in themselves good selling points, but when contrasted with the success, and ingenuity, and perseverance and triumph of the particular manufacturer of the product which is being promoted, then you have the beginnings of an interesting and convincing show.

For instance, one could take totally unrelated examples of good communication. The BBC programme '*Tomorrow's World*', the promotion of Skytrain and the film '*All The President's Men*'. The first is a weekly TV show about forthcoming products, the second is an enterprise offering cheap air travel and the third is a film showing two reporters mainly talking on the telephone. Yet it is drama which is the dominant feature of each example and which is largely responsible for their success. '*Tomorrow's World*' looks below the surface of modern technology, it shows why and how products are made. The emphasis is on development (in itself a dramatic element). Freddie Laker's well-publicised struggle in launching Skytrain has given a dimension to the service that he offers beyond that of the service itself. And the re-enactment of the Washington Post's Watergate investigation showed that great drama can be extracted from a situation that contains minimal physical activity. Detective work, analysis and deduction have enough drama to sustain a full length film or a novel. It is not unreasonable to assume that an audiovisual show can benefit from a similar use of drama.

Movement

The audiovisual scriptwriter sacrifices movement rather than drama. The images are still. Each one is read by the audience until the next image, or set of images, is submitted to them. But because these are projected for a short duration there is at least the possibility of evoking movement. A still image of a man reaching for a gun followed by an image of him pointing it at a protagonist, followed by another of his opponent outstretched on the floor will, with good timing evoke a sense of movement. The audience will fill in the spaces. This type of sequence is difficult to pull off, and it would be absurd to try to make the whole show in this style. A broader sense of movement can be achieved by showing distinctly differing locations and linking them with sufficient titles or graphics to prepare the audience for the change.

The only real movement which can be seen is the most basic form of animation where three of four positions are shown one after the other.

Again this must be used sparingly because it can appear too like a gimmick unless properly controlled. The reason for these animation effects appearing to be a gimmick is because in real life few activities are repeated mechanically in exactly the same way. The animated effects in multivision are best reserved for showing industrial processes, mechanical devices or in order to bring graphics to life.

Therefore, in multivision, movement is supplied by the audience, either imaginatively, or else simply by eye movement in reading the images or searching across a whole array of screens. In a large multivision show, eye movement becomes extreme, and the scriptwriter must ensure that the programme is paced so that the audience is not exhausted after the first two minutes. Rapidly changing images on each individual screen must be alternated with more restful panoramic shots.

Pace

Although a succession of still images, multivision is of quite a different order from individual colour photographs. What the audience sees depends on how long you show them a particular image. If, for instance, you show a table laden with food for $1\frac{1}{2}$ seconds, the idea will register with the audience: 'a table laden with food'. If you show it for 30 seconds, the audience will see a table with meat, vegetables, cutlery, a pink tablecloth, tall wine glasses, a bottle of white wine, etc. Some of the audience will begin to speculate why white wine is being served with red meat and others will question the colour of the tablecloth. If you were to show the image for, say, 6 seconds, the audience would begin to notice the individual items but would be forced to abandon this reading exercise as the image is replaced by another. Much depends on the pace of the programme. If all the slides are retained for a longish period, then the audience can gauge how long they have got before the picture is replaced.

Timing and pace are two elements for which the scriptwriter must have an instinctive feel. A stop watch is no substitute for understanding how a succession of images and sounds will come across to an audience. It can be learned by watching lots of shows and judging your own reactions and those of the audience. More effectively, it could be learned watching Woody Allen movies or seeing Bob Hope, Jack Benny or George Burns telling jokes. A former colleague of mine once suggested that I prepare a series of lectures on the theme that scriptwriting was like telling a good joke. I think the analogy would have been carried too far but it does no harm to bear that advice in mind while writing a script.

Essential ingredients

Tension, anticipation, surprise, reward, revelation—these abstractions are vital parts of the scriptwriter's craft. Not only must they be built in to a script but they must be carried through to the finished programme. They are also the ingredients which are often left out of audiovisual scripts. Perhaps writers think that they can be put in later—the director will say 'Put some music in here', or 'Chop the commentary there and go to the next sequence'. It is true that this can be done, but it is a rough and ready kind of butchery that is bound to affect the content of the whole programme.

The multivision medium

At the moment multivision lacks an identity and it will be several years before it has its own aesthetic. New multivision programmes continue to be experiments in form and it will be a long while before all the rules are made and even longer before they can be successfully broken.

Perhaps the emergence of audiovisual has stemmed from the changing aesthetics of film. It is the experiments of film-makers in rejecting the traditional forms of narrative that has encouraged this new medium. Particularly, the French New Wave of the 1950s and 1960s, in exploring new forms of expression, rearranging 'real' time and space in their films, created a visual language that was capable of stepping into another medium. Other directors, Visconti, Bergman, Losey, reduced action to a minimum. Antonioni said that he aimed to 'undramatise' in his films—true, up to a point, in that drama is taken away from action in order to highlight the drama of psychological conflict and photographic imagery. One New Wave director even made a film composed entirely of still images with only a few seconds of real movement. This was Chris Marker whose film 'La Jetée' brilliantly showed a future world under a nuclear catastrophe and its effect in a moment on a few participants. It could, as it stands, be a multivision show.

The unities of time and space are no longer essential for an audience to understand what is happening. People who have never heard of Alain Resnais will happily watch a TV play that ranges in and out of a chronological sequence. And to a critical viewer it will not appear disturbing—this is fortunate for the audiovisual writer because it means that his medium of still images and real-time dialogue and commentary can evoke an actual narrative which will hold the show together.

The content of the show

There are producers who believe strongly that content is all-important and that if the sponsor wants a show that sells, then that show must stick to the point from beginning to end; but of course a degree of subtlety is necessary in all selling, especially via a visual medium. If you have a captive audience for 10 or 20 minutes it should be possible to convince a high proportion of them that they really want a particular product. This is a discovery that the potential buyer must make for himself. Audiovisual gives the opportunity of selling in depth, of reassuring the customer that the processes of manufacture are *the* best anywhere, that the product is reliable, attractive and useful. The show must also anticipate all the customers' objections. Because so many audiovisual shows relate to marketing and selling, it is essential that the scriptwriter understands these activities.

Start at the beginning

Assuming that the scriptwriter has a full grasp of the subject for an audiovisual show—how does he begin? This varies from person to person, but nearly every writer begins with a broad outline—a framework of ideas on which to build the sequences. While researching the project, the writer will constantly bear in mind that he has to invent ideas for, say, a 20 minute show. The brief given to him should specify what message has to be conveyed and list essential information that must be contained in the script. Other optional information, general background, will be discovered during the period of research.

The importance of research cannot be overemphasized. The sponsor may well be very helpful in providing a whole stack of facts and figures but he may not include the kind of detail that stimulates some original ideas. For instance, he may say that a new factory is 60 000 sq. ft. in area but he fails to mention that the foreman is World Amateur Boxing Champion. This sort of information is a gift to the scriptwriter, because it can suggest all kinds of themes that would be both relevant and entertaining.

The scriptwriter makes notes that contain some potentially fruitful ideas. He selects the best of these and develops them. He also looks for a suitable style and pace for the programme, and actually puts himself in the frame of mind necessary to write in that particular style. Some writers approach every subject in the same manner, and impart a particular style to every programme they write. This can be of some advantage to a sponsor since he can easily guess what the result will be. However, the best writers are those who can genuinely adapt to the occasion, who can write humourously or seriously, high or relaxed tempo, using a whole variety of techniques.

The outline, or structure, emerges from the initial notes. This is an outline for the writer only, and it is not for submission to the sponsor. Its main purpose is to enable the writer to see that programme as a whole. It allows him as well to be sure that the programme fulfils the object within the budget. From the outline he will be able, from the start, to give the appropriate weight to each of the ideas that will be developed. It is easy to alter a one page outline but very difficult to change the shape of a twenty-five page script.

Develop the outline

The outline should be discussed with the director who will be able to bring a fresh mind to the subject. The photographers and artists may well be able to contribute at this stage. The writer can tell immediately whether their ideas are usable. Although he has written only an outline, he will have conceived the show as a whole. What he needs is more ideas in order to help it to grow.

Conceiving and developing a script is an exciting and slightly bizarre process—bizarre since it is not entirely logical. Reason, analysis, argument, will play a big part, but there comes a stage when it is more creative to pursue quite unrelated ideas. No doubt the sponsor would be surprised to see a man who is being paid a small fortune a *day* rummaging through old colour magazines, changing channels on the television and examining the contents of the refrigerator. But, in that case, refer him to Arthur Koestler's '*Act of Creation*' or to a book on lateral thinking and perhaps he will understand. The writer will be saturated with the information he has collected and he needs to look beyond it in order to find interesting and original themes. An original idea is like a magnet—suddenly, when a good magnet has been found, the important information to be contained in the show will cling to it like iron filings. When this happens the writer is earning his money because he is creating a programme rather than merely processing information.

A treatment

Many writers have worked on documentary films and it is usual to produce an intermediate document called a treatment. This is a lengthy description of all the sequences to be made, with indications of style, pace, location, commentary and visual and sound effects. This can be a redundant exercise involving a great deal of work to very little end. If you are accustomed to writing and reading finished scripts you quickly pick up the shorthand methods of expression. It becomes as easy to assimilate the columns of sound and vision so that you can imagine the show being played. In the same way a conductor will be able to 'hear' a piece of music by reading the lines of notation. It is therefore annoying to spell out everything in 'longhand' for the benefit of people who are unused to reading scripts. It is far better to go immediately to the script stage and to make sure that you are present to explain your ideas if the script is to be read by people who are unlikely to understand it.

The first draft

From the outline the writer can go immediately into producing a first draft script. This contains numbers indicating each change of slides. It has the commentary in full, together with sound effects, a brief description of the music to be used, and a description of each individual picture, or set of

126

pictures. Occasionally a writer may give only a brief outline for a particular sequence, for instance:

Picture	Sound
Sequence 10	
Six shots in rapid succession of	FX: JET ENGINES
modern British aircraft in flight	FADE FX AND BRING UP
	MUSIC 'MARCH OF THE RAF'
	FADE MUSIC

This allows the director to choose the pictures and form the sequence. There are a dozen different ways in which it could be realized. The aircraft could be stylized drawings, or real life shots. They could be arranged in any kind of visual pattern on the screen. Similarly equal weight is given in the script to the sound effects and to the music. The message of the show can be altered according to the direction of the material.

A good script will not be vague but it may well be very cryptic, leaving much to the imagination. In practice, this can work very well, allowing other people in the team to use their own creativity in making the finished programme. It is insulting to a photographer to say how a particular subject should be lit, but it is essential to tell him the details of the effect that you want to create.

The first draft, then, is almost the final script. Minor corrections may be made by the sponsor, more details put in for the final draft, but essentially this is the working document for the production team. Several copies are made, and a master copy kept by the director for noting any changes.

Presenting the script

The actual physical appearance of the script is very important. Since it will be read by the sponsor, perhaps passed around a boardroom table, it must be dressed for the occasion in a proper binding and be neatly taped and duplicated. Its job is not only to act as a design for an audiovisual production but also to advertise the expertise of the production team. If the items which it contains are sound and if the effects are described in such a way that even a layman can see that they are achievable then it cannot fail to win approval.

Numbering The script should be numbered with cues indicating each slide change. It is not necessary to write in every single dissolve, cut or snap change of image. This makes it more difficult to read and is best reserved for the later, cue sheet, stage.

Sequence numbers or section numbers are useful, so that anyone can quickly refer to a particular part of the script. It is easier to remember 'sequence six' rather than cue numbers 'one hundred and twenty-two to one hundred and forty-five'.

Directions   All directions in the 'sound' column should be in upper case letters in order to distinguish them from the spoken commentary or dialogue. Actors and commentators are familiar with this convention from TV and film work.

Typing   TV and film scripts are laid out so that dialogue is heavily indented and forms almost a separate column down the right hand side of the page. The description of events overlaps this column but is not adjacent to it. An audiovisual script benefits from the earlier convention of having two columns side by side. It is not that the medium is more reliant on split-second cueing but rather that it is more difficult to appreciate the cueing when reading the script if you have to read a picture description followed by the accompanying commentary.

The typing should be well-spaced, and this in itself should help to indicate the pace of the programme. Larger gaps can be left between sequences, and so on.

Content   Everything in the script must be a means to an end—the finished programme. All unnecessary decoration should be omitted in the final typed script. Unless the information is going to be used, it is useless and even misleading. For instance, if the show contains a character who is not referred to by name, it is no use calling him 'Sir Montague de Beavoir' or anything other than a cryptic DG (for Distinguished Gentleman). Otherwise the script is being written for a reader rather than as a means to the programme.

MULTIVISION STORYBOARD	PAGE
TITLE	DATE

CUE	VISUALS				KEYBOARD	TIME	SOUNDTRACK
					□□□ □ / □□□ □		
					□□□ □ / □□□ □		
					□□□ □ / □□□ □		
					□□□ □ / □□□ □		
					□□□ □ / □□□ □		
					□□□ □ / □□□ □		
					□□□ □ / □□□ □		
					□□□ □ / □□□ □		
					□□□ □ / □□□ □		

Storyboard.

128

The storyboard

When the script has been finished, the scriptwriter still has work to do. The production of the storyboard is the next stage. This is a series of visual notes, jottings which indicate slides, and which will ensure the visual continuity of the show. A storyboard is essential in any multiscreen production. In fact, it may be decided that a script cannot physically accommodate the appropriate visual directions that are necessary. If you have to refer to specific images on each of six or twelve screens, then a storyboard with the screen matrix drawn on the left hand side and the sound column on the right will be easily understood by everybody. There is a danger in going straight to the storyboard stage in that undue emphasis will be given to the visuals at the expense of the soundtrack. For effective presentations, sound and vision should be conceived together to form a single audiovisual image. In practice it is frequently the soundtrack, particularly a commentary that carries the show forward and gives it continuity. The commentary is easily understood while the eye struggles with a succession of visual images.

The soundtrack

The soundtrack should be harder to understand; in efforts for clarity and communication, the soundtrack walks along at a pedestrian pace, often stating the obvious. In 95% of programmes the soundtrack would benefit by demanding more attention from the audience. A busy soundtrack with plenty of effects and a variety of voices and music will help to balance the vast amount of visual information that is usually provided.

This common fault stems from production methods and from the actual visual bias of audiovisual creators. Sound recording has been with us for a long time and it is not as new and exciting as all the technology for projecting pictures. But more fundamental than this is the custom of recording a commentary in an arid room, with a chair, table, microphone and glass of water. This is scarcely the world in microcosm! And from this beginning one adds music, also recorded under clinical conditions—and a few stock sound effects. Whatever its technical merits the soundtrack will be like a monochrome jig-saw puzzle when compared to the vast Impressionist painting of the visuals. The photographer, by contrast, will have ventured out into the real world and however selective he has been he will have recorded a version of reality and a thousand details and accidental effects with which to assault our eyes.

In making an animated film one puts together the soundtrack first of all; it is the basis of the film. I remember visiting an animator at his home and was shown into the living room where he had at least six tape recorders, a sound mixer, several headsets and assorted record decks, splicers and tapes—but he had to hunt round for a pencil and paper! The audiovisual industry would benefit from an influx of good animators who understand the importance of sound.

Visual effect

It would be helpful, too, if an enterprising producer experimented with visual images and their effect upon an audience when accompanied by varying sound images. A precedent would be Pudovkin's experiments in montage which helped to build a film language now used every day in commercials, features and TV programmes. He intercut a close up of an actor first with shots of food, then with an open coffin, and again with a shot of a girl playing with a toy bear. In each case the actor appeared to be relating and reacting to the particular image. Even his face seemed to register hunger, grief, and amusement. And yet, Pudovkin used the same piece of film in each sequence.

Another example, again a purely visual exercise, is the brilliant way in which some film-makers can control the environment in which they set their actors. A person photographed against a prison wall actually appears to be in a different state of mind from someone photographed in a garden. With these phenomena being subconsciously appreciated by everybody it is surprising that many people engaged on audiovisual production do not attempt to control the impact of the combined audiovisual image.

Work on the storyboard

However large the visual extravaganza is intended to be, equal weight should be given on the storyboard to the soundtrack. Some production companies will produce two storyboards: a rough storyboard for use in the studio; and a presentation storyboard for submission to the client. The first one would be sketched in felt tip pen and the second made in full colour, mounted on card and bound.

As the storyboard progresses it clearly demonstrates any faults there are in the mental visualization of the script. In multivision the scriptwriter/designer has to cope not only with continuity of images but also ensure that images on adjacent screens produce the right effect. In a large multiscreen it is usually preferable to have harmonious images appearing simultaneously. If five screens show strong colours they will tend to obliterate a sixth image which has less impact. This is where the storyboard is a great help in clarifying the visual design of the show.

The layout of the storyboard shows a new set of pictures each time there is a visual change. The numbers on the storyboard should therefore be cue numbers, corresponding to those on the original script. Now, cues will not be equally spaced and there is a danger that the indications of timing will be lost when the script is turned into a storyboard. This is another reason

130

for including all the commentary, sound effects and music directions. It is, after all, a storyboard and not simply a 'pictureboard'.

This can be partly overcome by writing in the time next to the cue number. A further indication can be made by writing—in the screen duration of each image. It would be preferable, but not graphically possible within the storyboard to indicate the timing in visual terms. This can be more easily achieved on a cue sheet, which is a later stage of production.

Plotting tension

A useful exercise for animators, recommended by John Halas is the 'tension chart'. This is drawn up before any animation is begun. It could well be used by multivision producers. Its function is to show the pace of the programme in graphic terms. For a multivision show it could plot image changes against a horizontal time scale, and against a vertical 'pace' scale that shows dissolve speeds. A simplified version which plots image changes against time in seconds may be easier to read and could be included at the foot of each page in the storyboard.

Pace is affected by the amount of information that is being displayed on the screen. At the beginning of the show the unfamiliarity of the particular format can dictate a relatively slow pace. As the audience becomes accustomed to the location of the screens, the pace can quicken and the information is accepted.

The advantage of a tension chart early in production is that it shows whether the pace of the visuals is in keeping with the soundtrack, and with the intentions of the producer. Intelligent use of it indicates whether more slides are needed in a particular sequence—and it will do so before the photographer has handed in his work and gone home.

If you intend to draw up and use a tension chart you should ensure that a change of image really does constitute an addition of information. For instance, if twenty screens change one after the other, but they all show

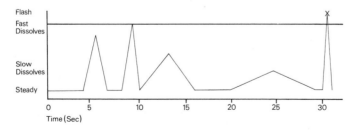

Tension chart for audiovisual shows.

aspects of the same scene, it is likely that the audience needs only a short while to assimilate it. In spite of the rapid changes there is no tension. You should therefore examine the chart realizing that the flow of information is not the same as the flow of images.

Of course individual reaction time varies. One way around this is to include little details to amuse the more quick-witted in the audience: signs, visual puns or cunning superimpositions of minor detail. If someone misses it they will not be missing the main message of the show. These details add to the professionalism of the craft and the connoisseur will note them with appreciation.

The storyboard shows the counterpoint between the visuals and the soundtrack. For instance, key words in the commentary should not coincide with slide changes, or they will be lost. However, the main function of the gigantic 'cartoon document' – which can run to many pages, each of them A3 size (or larger) is to begin to separate the project into its component parts. What began with the outline and the script as a flow of audiovisual images now becomes a series of instructions to the photographers and artists. But however skilfully put together the storyboard is only a framework and requires constant interpretation by the director.

At the beginning of this chapter I said that a script must be able to create an exact impression of the show in the mind of a reader. Although it will describe the show in terms of specific audio and visual images it will not contain exact instructions of how to achieve them. The storyboard is the analytical stage, the first practical step in the production of the programme.

8　Graphics

The visual source material for a multivision show may be either 'real-life' photography or specially drawn and prepared artwork. Multivision is a graphics medium, whether the pictures are *photo*-graphics or *art*-graphics. Therefore, these two sources should not be considered as totally separate. In fact, the processes which are now used to produce the first set of slides tend to bring both extremes, recorded reality and graphic representation, into a middle area. Photographs of people, objects and scenes are processed to make the pictures more graphic—and drawn artwork is animated in order to bring it to life.

This evolution in the production of visuals for multivision shows is the result of experiments by producers working in the medium who are aware of techniques which have been developed in other media. Print production and packaging, film animation, television graphics; all of these are rewarding fields of study to anyone interested in designing for audiovisual. They are more relevant than the traditional media of documentary photography and film production, although even these have techniques which can be used.

Equipment manufacturers have responded to the growing needs of multivision producers by designing equipment specially for use in complex slide preparation. The biggest advance has been in the design of rostrum cameras and in duplication systems which make the whole process of slide presentation a precise and exacting craft—as well as opening up thousands of artistic possibilities. The operation of the rostrum camera is considered in detail later, here we consider the rules and methods of artwork production, remembering that the technology and chemistry of photographic processes are integral parts of the work.

Audiovisual requirements

The production of slides for audiovisual aids is an art in itself, but the end-product cannot necessarily be used in an audiovisual show. Even the most commonplace single-screen programme should attempt more than merely showing a series of graphics accompanied by a spoken explanation of them. In an audiovisual show the graphics should speak for themselves. Not that they should be verbal graphics; on the contrary, these should be kept to an absolute minimum. But a sequence of slides showing artwork-based material should be intelligible with a minimum of spoken explanation.

Simplicity is the key to good artwork; not only should the design concept be simple but also the execution of it. If material is being produced for a brochure or any form of printed medium then it can have a wealth of detail with which to communicate information. But a slide appearing on a screen for a few seconds has only a short space of time in which to impart its message.

There is a further difference between printed material, particularly advertising matter, and audiovisual communication. An audiovisual show does not have to compete for attention. It will probably have a captive audience who are able to concentrate on whatever you are to show them. Even a simple, unobtrusive image, which would be lost as a poster in the market place, can have great impact in a show. While the technical processes that are used in advertising, packaging and print production are extremely useful to the multivision designer, he must be careful not to imitate their form and content.

It is far better that the freshness of the original design idea should be seen on the screen than for a particular drawing to be meticulously coloured and shaded as though it were intended for hanging in an art gallery. Some bad advice was once given by a lecturer at a film school: every frame of a film should 'be a Rembrandt'. By this he meant that, wherever the camera pointed, the subject should be perfectly lit, the image perfectly composed and the exposure completely accurate. And just as you cannot select a single frame of a film and expect it to bear scrutiny as a finished work in itself, so you cannot view a slide from an audiovisual show out of context.

Remember the context

Unfortunately, artwork has to be produced 'out of context' in the studio. As each piece is worked on individually there is a natural tendency for an artist to decorate it until it looks pleasing on the drawing board. However, experience will show him that the projected image has quite a different effect. In a big production it is essential to immediately photograph the first pieces of artwork and to see how they appear when projected. In this way it is easy to see how the finished sequence will look. It might be found, for instance, that a subtle pastel shade that looks fine on paper appears dull and sickly

In this installation, the multivision images were projected onto 4 m (13 ft) diameter spheres, with three or four pairs of Kodak Carousel S-AV 2000 projectors for each sphere. The projectors were suspended just below the ceiling.

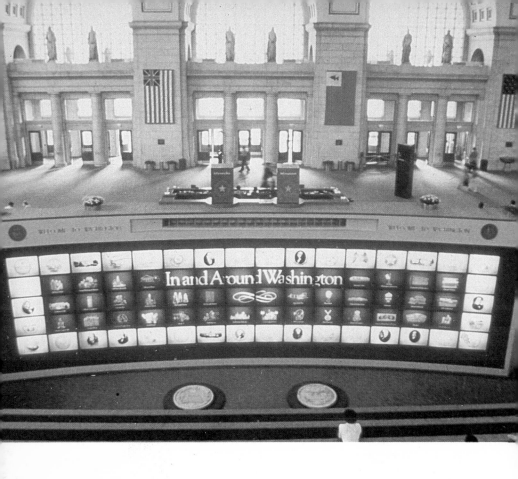

Above.   The massive 80 screen multivision installed at the Washington National Visitor's Centre in Washington, DC, USA. This was previously the Grand Union Railroad station and as you can see, it required a very large multivision to have any impact on the huge environment. The complete installation was carried out by Atlantic Audiovisual of Montreal and New York.

Opposite.   One of the most comprehensive multivision presentations was that designed for the 6th Bahman Museum in Tehran. The layout involved movable columns of screens (*top*). One projector was aligned with the auditorium seats (*bottom*).

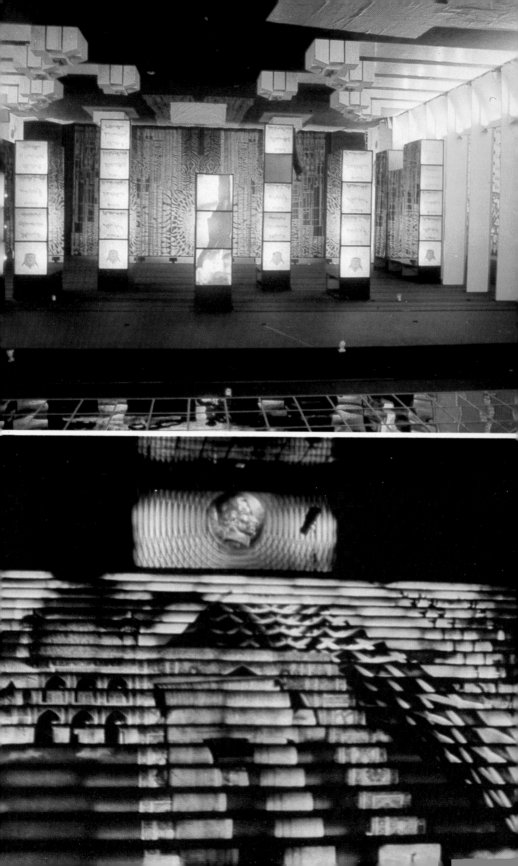

St. Michael's church in Chester. This church which is architecturally interesting but it is no longer used for worship, has been turned into a city visitor's centre and is called the Chester Heritage Centre. This includes a 3-screen multivision which describes both the city's past and the planning decisions that have to be taken for its future.

To provide an added attraction to Madame Tussaud's space tableau, images were projected from floor level on to the moonfarers' helmets.

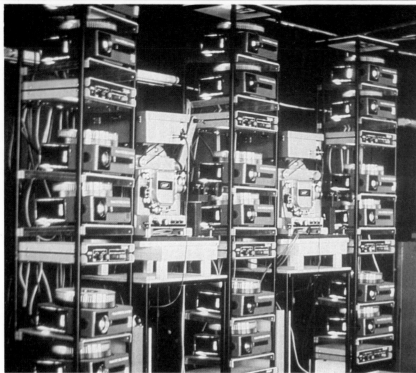

Above. The mosaic multivision can still be one of the most effective uses of the multivision medium when done. A recent example is the 24-screen used by Lucas which has been seen in Detroit, Tokyo, Geneva, Frankfurt, Paris and England. The lower picture shows the bank of projection equipment used. The show is a prestige presentation—very much to say we are the biggest and the best.

Opposite. An exhibit in the Evoluon Eindhoven, called the 'communications theatre'— it is the audiovisual story of how man communicates. Each of the 'building bricks' is about 60 cm cube and is fitted with a compressed air operated flap front door. Most of the cubes have inside a rear projection screen which uses either slides or 16 mm movie. In all there are about 7 slide screens, 2 movie screens, 7 lighting circuits and 25 mechanically operated items. The complete system and programmes are shown alternatively in English and Dutch. Designed by the British designer James Gardner, it includes many sound and light sequences and audiovisual displays.

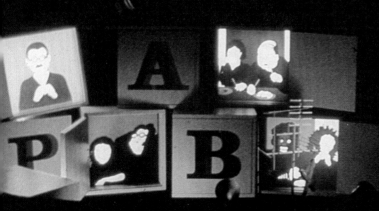

Six-screen presentation room in Norway shows an ideal small-scale layout for intimate presentations.

on the screen. It would be disastrous to produce the whole show in such a style.

A great help during production is to have a complete copy of the storyboard pinned up in the studio for reference. This serves as a constant reminder that the end-product is an audiovisual show rather than a finished piece of artwork. The presence of the storyboard helps the continuity of the show by showing the context of each picture. And its 'shorthand style' indicates that the simplest design can communicate effectively.

Artwork categories

So what are the main categories of artwork which the artist will be expected to produce for an audiovisual show? They will certainly include lettering, freely drawn cartoons, charts, maps, flow diagrams, abstractions of industrial processes, composites using cut-outs—in fact a wide range that require different skills. It is usual for an artist to develop his expertise in one or two areas so that people who commission him know what to expect. Really gifted artists can find themselves out of work because their portfolio is too varied to make a good impression. And not only do artists specialize in particular styles or techniques but many of them concentrate on certain types of subject matter: botanical drawing, figure drawing, horses, boats, motor cars and so on. That is fine for printed material, but audiovisual studios should employ the more versatile type of artist—the jack-of-all-trades who, hopefully, can learn to be a master in the audiovisual medium.

The exception to this may be the artists who specialize in titles, captions and lettering. Few cartoonists have the patience to produce good lettering and even in a multivision show the smallest imperfection in a typeface will be apparent.

Lettering

Lettering can be used by itself in a single slide, or it can be photographically combined with another image, or it can be superimposed during the show from another projector. Whatever method is used, the words must be instantly readable. This rules out many standard typefaces that are used in printed material. However, there is wider scope for unusual lettering than there is in television where the screen raster tends to obliterate horizontal lines and serifs. (Of course, if a sponsor insists on transferring a single-screen show to video cassette then the problem is still there.) On the whole, tall, thin letters should be avoided, being more suitable for the shape of a newspaper than for the 3:2 ratio of 35mm projection. Highly decorative styles are also unsuitable because there is no time to appreciate their detail.

Fortunately, lettering in audiovisual shows does not have to be dull. There is more scope for interesting titles and captions than in any other medium. This is because the graphics can be very simply animated, made to grow bigger or smaller, rotate or jump up and down. They can change colour or appear to be composed of bricks or steel, grass or water. These effects are possible because of the processes of registration photography, colour filtration and photographic superimposition.

It is likely that the person making the most contribution to the graphics in this case will be the process photographer at the rostrum camera rather than the artist. In fact there is nothing to stop anybody from using white rub-down lettering on a sheet of black card and sending it to the studio with instructions for a really complex sequence to be photographed. The photographer will be able to zoom in on the word, select individual letters or reveal them one at a time, add any shade of colour to either the word or the background, and superimpose patterns and tones. The only difficulty is in first composing the written instructions in such a way that the end result is what you wanted.

The artist, therefore, is concerned with designing sequences of lettering although the physical work of producing it is quite mechanical. There are around 700 styles and sizes of rub-down letterings available. It might be thought that it is not necessary to worry too much about the size of lettering since this can be adjusted on the rostrum. To some extent this is true and successful sequences have been made using all sizes of artwork. However, it is sensible to establish contact with the photographic studio and find out what they prefer. They might, for instance, have installed a new Oxberry Pro Copy F-2 rostrum which has a slide copy field guide. This establishes a standard size of artwork $25 \cdot 8 \times 30 \cdot 8$ cm ($10\frac{1}{4} \times 12\frac{1}{4}$ in) and it indicates a number of slide formats (Wess mounts 2, 9, 16 and 34). The artist has a choice of using clear cels or bond paper, depending whether composite images are built-up or if a single piece of artwork is being photographed. The use of this type of equipment is essential for good registration work and consequently to all audiovisual graphics. A guide to artwork sizes is set out at the end of this chapter.

Adding animation

It is with the more complex registration effects that the techniques of the film animator are called upon. The ability of audiovisual to show graphic representation of industrial processes is almost equal to that of animated film. The very fact that a diagram is built up in stages is an advantage. It is easier for a viewer to remember, say, the design of a particular machine by looking at a number of simplified abstractions rather than by seeing moving images of the same machine. In any case, an animated film of such a subject is unlikely to use a frame-by-frame type of animation, and so the two styles are not dissimilar. However, the cost savings of using audiovisual for this subject are enormous.

136

Such a sequence is produced by drawing each stage on clear acetate cels. The cels superimposed one over the other show the complete picture. The cels are photographed one at a time or in combination on the rostrum where they are held in exact registration by a peg bar. If an unchanging background is being used, then this can be painted on strong paper which is also punched with registration holes to keep it in position. The cels should be numbered so that the cameraman knows in which order to photograph them.

Backgrounds

The background on any graphic slide should be dark in tone. The important information will then stand out more clearly. The letterings should normally be lighter than the ground.

If lettering is black on white when it is projected, the white background will dominate the screen and temporarily blind the audience. Not only that, but specks of dust will show up more clearly and will be magnified by the projection lens to enormous proportions. It is always difficult to get rid of all the dust since the projector fans draw it in with the cool air. But the effect of it can be reduced by using sensible design in graphics. 'Real-life' photography will tend to have more detail, more areas of light and shade, rather than flat one-colour backgrounds. Dust specks will not show as clearly on this type of picture; but graphics require more care.

Maps

All of the 'rules of thumb' which I have mentioned above: the need for simplicity, for readable lettering and for dark backgrounds, apply to the drawing of maps for projection. A cartographer's map is something you study closely for a long period. But an audiovisual show requires a map which has specific information, for example, the relative positions of three towns. In this case, a map is easily produced by tracing a simplified coastline onto a sheet of card, cutting around it and pasting it onto another sheet of a different colour say, blue, for the sea. The names of the towns can then be lettered either directly onto the card or onto separate cels if bump-on names are required. If the land-mass is in a dark colour the words could alternatively be superimposed from another projector.

Sequences

A series or sequence of pictures is almost always better than a single picture. This is normally indicated by the writer, but an artist who knows the nature of the medium can make helpful suggestions. For instance, the cut-away technique is a good teaching method and is used frequently to show the interior of machinery or even biological specimens. The first slide shows a normal view of the subject, while the second slide shows a cross-section, in a perfectly registered position. This technique may be taken further by making the initial picture an actual photograph of the object. This would be followed by a graphic drawing of the object in the same position—necessary to bridge the transition from 'real-life' to graphics. The third slide would show the cross-section. There is, of course, more work involved in producing a series of graphics and it will use up the slide capacity of the projectors. But from an instructional and aesthetic point of view it is always to be preferred.

Exploded diagrams are a good way of showing how a piece of machinery is assembled. Frequently, however, they can contain too much information for inclusion in a projected show. If a specific point is being made, such as the function of an individual part, then this can be highlighted by giving it a colour while retaining grey tones for the rest of the picture. Again, using a second slide to show the additional colour is even more effective.

Graphic style

Although the use of diagrams is closely associated with teaching programmes, audiovisual techniques enable an artist to reach a different kind of audience. In advertising presentations, for instance, there is an opportunity to use really stylish graphics in keeping with the style of the programme. A visual theme can often be taken from existing material, for instance, the design of a logo or the shape of a package, and developed for use in a short sequence. A really complex animated sequence could be worked out for use in several parts of the programme as a linking theme. The extra cost and time involved can then be justified.

The visual style of the graphics and illustrations used in an audiovisual show should be consistent throughout. On a large production the team of artists has to be organized so that individual styles do not conflict. Because of the need for speed a studio has to use factory methods of production with one person responsible for the overall design and other fill-in artists doing the more menial work. This is not a bad thing in itself and the artist may take comfort in the fact that Raphael worked in exactly the same way.

The style of the show is an integral part of its message and it must be one of the chief concerns of the producer, writer and artist. Certainly the style has constraints placed upon it, such as budget or time limitations, but it is not really linked to anything other than the flair and inventiveness of its creators. Multivision is a highly technological medium and it is sometimes necessary to humanize it by using a more relaxed style of graphics. On the other hand it may be argued that hard-edge graphics are sympathetic to the photographic nature of the medium, or that a wealth of air-brushed artwork is preferable to a sketchy treatment with plainly visible brush strokes and pencil outlines. It really comes down to personal opinion and to the taste of the individual who views it.

138

Strip cartoons

Finally consider the use of cartoon illustration in the multivision medium. Their two-dimensional form is suited both to the medium and to the production methods that are used. Cartoons can distil the essential points of a message by a clever use of cliché and caricature, and there is no reason why a whole show should not be based on them. Cartoon characters can link a show together in a way in which still photographs of real people cannot. Although I think there is far more scope than is realized for using the narrative form with real people being shown, the absence of movement is felt more strongly. This is because a still photograph takes the subject matter *out of time* and the multivision medium puts it back into its *own* time. The resulting effect is that the characters cannot develop, however many pictures you show of them, because they have no means of pulling the audience into their own rhythm. Strip cartoon characters are completely artificial and we do not expect to have a real-life rhythm. But because they are cliches we automatically 'add on' all the other characteristics, including the ability to live in time.

Strip cartoons in print, especially in science fiction magazines, have often overcompensated for the lack of movement in the page. Extreme uses of diagonals and perspective add almost another dimension to them. In other cartoons, e.g. Charlie Brown of '*Peanuts*', the static quality itself is caricatured. Now, if you add the time flow of audiovisual to each of these styles, what happens? In fact, they both work because the all-action strip cartoon actually shows a moment of extreme activity, a peak dramatic point, rather than a process. The ultrastatic cartoon, too, is effective because accompanying words on the soundtrack can show what the character is thinking.

A guideline to the scriptwriter is that he should make the cartoon characters appear in situations where they talk about what they are going to do or what they have already done, rather than showing them engaged in a particular activity. But an extreme dramatic moment can be equally effective because the audience will imaginatively fill-in the sequence of events leading up to it. Observe also the irony that cartoon characters appear more alive in audiovisual shows than real people.

Artwork sizes

Standardization of artwork sizes is of enormous value in speeding up the work of the studio. It can help the artist, the photographer and, especially, the people who have to file it after the production is finished.

The recommended standard for artwork is 25×30 cm (10×12 in). This is the paper size. Inset into this rectangle is the actual information area that will conform to the screen format. For 35 mm slides this will be $15 \times 22 \cdot 5$ cm (6×9 in). It is essential to overlap the exact information area by approximately $1 \cdot 25$–$2 \cdot 5$ cm ($\frac{1}{2}$–1 in) which will allow for line up on the rostrum.

Therefore, on a 25 × 30 cm (10 × 12 in) paper the picture will be approximately 18·5 × 25 cm (7 × 10 in), allowing a margin for instructions, registration holes, attachment of cels or simply for safe handling. The basic standard extends to other formats.

TV 'safe title area'

It is to be hoped that the programme will not be transferred to videocassette —but in case a particular slide is destined for TV viewing, the figure shows the standard 'safe title area' for a 35 mm slide.

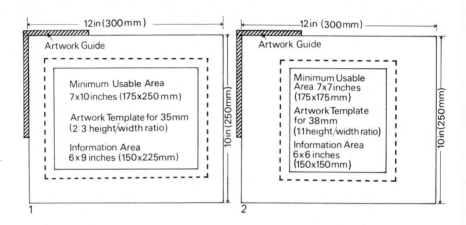

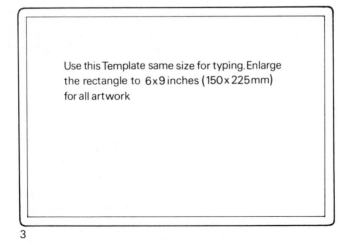

Artwork sizes. Artwork templates, 1 :35 mm; 2 :38 mm.

Lettering sizes—general guide

Once standard sizes have been applied to the artwork it is then possible to apply standards to the *minimum* size of lettering used. One of the most common mistakes in audiovisual shows is to make the lettering too small, and hence illegible to people in the back row of seats.

Later in the Designer's Guide, recommended sizes of screens in relation to room and audience size are given. Assuming that these are observed, the minimum letter height should be $\frac{1}{50}$ of the information area. This is the absolute minimum and about *double* that size is more comfortable to read. The letter height is taken to be the smaller dimensions of lower case lettering (i.e. excluding the tails on b's, g's, etc.).

Given the basic standard for 35 mm to be $15 \times 22 \cdot 5$ cm (6×9 in) (information area), the minimum comfortable letter height is 6 mm ($\frac{1}{4}$ in). Unusual typefaces will not conform to this basic guide.

Lettering sizes—the academic approach

Academic research has been carried out to find the optimum sizes of lettering for projection—but the results are somewhat inconclusive because of the many variables in viewing conditions, typefaces, emulsions and the eyesight of the audience.

Take, for instance, the character height to artwork ratio. The Ergonomics Society think this should be 1:17, the Transportation Research Board (1974) 1:32, Kodak Ltd. 1:50. On standard sized artwork the lettering would then vary in height between 4 mm and 16 mm. How do the academies arrive at these varying conclusions?

The following formula was provided by Peters and Adams (1959), (quoted by McCormick 1976 and requoted by Galer 1978):

$$H = 0 \cdot 0022D + K$$

where H is the character height of letters (mm) on screen; D is the viewing distance; and K is a 'constant' whose value is determined by the viewing conditions and the illumination of the slide; a 'variable' constant. The smallest value quoted for it is $1 \cdot 524$, that is, very good conditions, and the largest is $6 \cdot 604$ (very poor conditions). Thus if a viewer sees a slide from a distance of $9 \cdot 75$ m with good viewing conditions the formula tells us that the letters should be 23 mm high on the screen. The letters will then subtend an angle of $8 \cdot 1$ minutes. Others have suggested angles of 18 minutes for words and, for some unexplained reason, 21 minutes for digits.

This obtuse approach to audiovisual is unrewarding and unrealistic, follow the general guide above.

Typewritten lettering

Typewritten copy is sometimes essential for speed of production. It is not ideal aesthetically, although expanded typewriter lettering can look very attractive if reproduced very large and with good contrast.

The minimum legibility requirements demand that height on the screen is at least a $\frac{1}{50}$th, or better, $\frac{1}{25}$th of the screen height. A useful standard, therefore, is to reduce the information area for 35 mm slides to 7.5×10 cm (4×3 in). Typewritten copy *cannot* be photographed from the standard 15×22.5 cm (6×9 in) information area. This recommendation includes all normal typewriter faces such as pica and elite. It does not apply to production of slides for TV viewing. Copy should be at least double spaced and limited to 8 or 9 lines maximum. The use of written copy, typed or otherwise, should be kept to a minimum.

Computer-generated graphics

Advances in computer technology have provided audiovisual producers with a new service: computer-generated graphic slides. These can be produced with a high line resolution, making them virtually indistinguishable from artwork-originated slides. As a service, the greatest benefit of this technique is the speed at which a slide can be produced. A company who has developed one process for doing this are General Electric, who are able to offer (at the time of writing) a two-day service for producing, say, 50 original slides. The technology consists of an operator's console, a mini-computer and a film recorder. Up to ten visuals an hour can be created, composed and photographed with this equipment. The computer memory has a store of images, symbols, letters, figures and other basic graphics, any of which can be rapidly displayed on the video console. The size and position of these images can be adjusted by a skilled operator, perspective can be changed and an almost unlimited spectrum of colours, shades and hues can be fed into selected parts of the picture.

The television display is a standard 525-line system although the computer holds an image which has a far higher resolution. Thus the image on the screen is a 'working rough' and not the final product. The console has a keyboard with a selection of switches, joysticks and LED display. A copy machine, linked to the console, enables the operator to print out black-and-white copies of the monitor image. These serve as a reference until the slides are made.

The computer has a disc memory which contains the store of information about symbols, letters and so on. The DEC digital type machines provide a store on which image data can be recorded and read. The system stores up to 56 images on one of the 4-inch DEC tape reels. These are stored as codes rather than as video signals. This information storage method enables a client to return at a later date and to call up the old images and up-date them.

142

It is not yet possible to scan existing photographs. Symbols and letters have to be entered in code which is a major task. However, in the future this is certainly a possible development.

The photographic system used in the chain consists of a recorder and two standard cameras, one movie, one slide (35 mm). The recorder uses a 5-inch cathode ray tube with a white phosphor. This gives a resolution of 4000 lines (compared to the 525 lines on the monitor). The images that can be taken from the tube are close to the resolving power of the photographic emulsion.

Colour is provided by a separate element: a colour wheel that spins in front of the cathode ray tube at 1800 rpm. The sequence of primary colours is repeated on the wheel so that each primary colour passes by 3600 times a minute. The flying spot scan of the recorder traverses the CRT, controlled by the computer. The spot flashes on and off to expose the correct amount or red, yellow or blue to produce the required colour and intensity.

It takes about 1 minute to expose an average composition, although some particular complex slides with a great amount of detail and colour complexity have taken up to 15 minutes to expose. In comparison with the hours that are spent by artists on drawing and colouring artwork this speed, which is considered slow for electronics, is extremely rapid.

A computer graphic console is operated by a graphics artist rather than by an engineer. True, he has to be skilled in computer programming, but primarily the skills required are those of an artist who can translate the user's requirement into material form. The highly developed technology associated with computer graphic production at present takes the work away from the in-house home base. It is possible that in the future reasonably priced equipment may be manufactured for purchase by the user rather than for the provision of a service.

Registration photography

Artwork that is to form a sequence of exactly registering images is usually prepared on clear acetate cels. These are purchased without registration holes punched into them. The artist may include a background image on which the cels are placed. They are collated by the photographer who uses a cel punch to fix their position relative to each other. Under the camera the cels are held in position by a peg bar and successive images can be copied.

The copy stand

A floor-standing copy stand is a very solid piece of equipment. It has to be, in order to eliminate vibration. Its steel column will be around 3 m (10 ft) high and about 15 cm (6 in) square. Because of the heavy weight of the camera this is counter-balanced by an internal weight. The camera is raised and

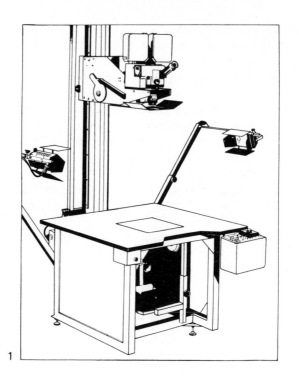

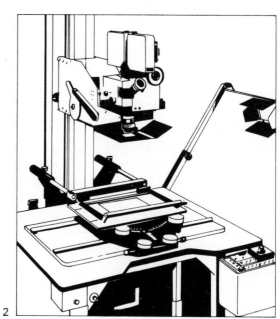

1, Fixed pin registration rostrum camera. 2, Registration camera with compound animation table.

by an electric motor, the speed of which can be varied by remote control. It provides smooth tracking with automatic cut-off of the motor as the camera reaches the upper or lower limits of travel.

A field size indicator is a useful guide to the cameraman. This can be a simple strip of write-on material mounted along the side of the column. The field size is shown by a pointer mounted on the camera. A versatile indicator is particularly useful, showing the field size for several gauges of film, say 16mm, 35mm and 46mm, and for each of the lenses used, say, 75mm, 55mm and 28mm. An alternative method is to have a mechanism built in to the column which gives a read-out of the field size on a digital counter.

The copy table

The basic support table is of equally solid construction with a working area 120×105 cm ($4 \times 3\frac{1}{2}$ ft). The notched design is now almost standard. A cut-out portion enables the operator to sit close to the work and yet the overall size means that it can accommodate very large artwork when necessary.

An illuminated aperture is provided in the centre of the table for lighting transparencies. This is about 35cm (14in) square with an opal glass insert for diffusing the light. Colour filters are normally mounted underneath the table top where one can position a diallable filtration colour head (as described in the section on duplicating). Cooling for the filters is provided by a blower mounted under the table and ducted between glass inserts.

The illumination system is balanced for 3200K, colour temperature, using tungsten halogen lamps. Because a large area has to be illuminated the power of the lamps is considerably higher than that used in smaller desk-top duplicators. The bottom light is about 1000 watts and has an automatic dimmer to reduce brightness after an exposure. Top lights are also provided, on adjustable brackets connected to the main frame. These are about 500 watts each. The illumination from the top lights may be polarized to give trouble-free shooting when artwork is covered by glass, or to prevent reflections from the surfaces of acetate cels. Polarizing filters have to be placed in holders in front of the lamps. Barn-door flaps make these lights directional, stopping stray reflections.

The very size of the console solves some of the constant problems which are encountered in copying. Vibration is eliminated and the heat problem is also reduced because the bottom light can be mounted near the floor, a good distance away from the transparencies.

Other problems, however, are created. It is difficult to focus a camera if it is 3m (10ft) up in the air. So, as the camera is raised or lowered a cam is turned which moves the lens. The cam is shaped to match the focus curve of the lens. This is a very critical part of the equipment and each individual lens is matched with a cam in the factory. Therefore the lenses which are used should be the ones already provided with the equipment.

Common to all animation stands is the shadow-board mounted beneath the lens on the camera mount. This is to prevent the camera from 'seeing' a reflection of itself in the material being copied, and also to stop reflections from the camera undercarriage.

The artwork table

Since the camera is providing accuracy to within, plus or minus, ·075 mm (·005 in) you must position artwork with the same degree of accuracy. This can be accomplished only by using an animation compound table. The compound table provides circular rotation through 360°, and horizontal movements: called east/west and north/south. The movements are achieved by turning callibrated wheels connected to counters which give readings to the nearest ·025 mm (·001 in).

Vertical movement for zoom effects is provided by raising and lowering the camera. A diagonal zoom is complicated and involves moving both the E/W and N/S assemblies as well as moving the camera for each shot. Complex movements can be more easily made if a pantograph table is attached. This is a table attached to the side of the main compound. A pointer, fixed to the compound and aligned to its centre, traces the movements on a piece of paper on the pantograph table.

The pantograph is useful in shooting off-centre zooms because a key plan can be traced from the artwork and placed on the pantograph table instead of using graph paper. The key plan must be placed 'upside down' in order to match the inverted image in the camera.

The compound table allows light to be projected from the bottom lamp in the console. The diffusion plate should allow full sized artwork, 25 × 30 cm (10 × 12 in), to be used. Compound tables vary in the degree of movement which they allow. At least 20 cm (8 in) of E/W N/S travel is needed. For movie animation a longer E/W travel is needed to provide moving backgrounds.

The peg bars for registering the artwork can be either static or moveable. They move with the compound, and may be moved independently.

Alternative kinds of peg bars are employed. There is the tape-on peg plate which can be mounted on any working surface. This has a wider base plate and will not fit into a standard peg track. On the most sophisticated rostrums a floating peg bar gives completely independent E/W movement to part of the composite artwork. This would enable an animator to create effects such as a car driving into frame with the background static. When the car reaches the centre of frame the background starts to move, creating the impression that the camera is tracking alongside the car. There is limited use for this degree of complexity when using still pictures and this facility is required only for full film animation.

146

Finally, on the top of the compound table something is needed to hold the artwork flat. This is the glass platen which swings up to accept the artwork and swings down to keep it in position. It must not touch the cels or transparencies otherwise Newton's Rings appear. The platen should accommodate different thicknesses of artwork, and even small objects such as books.

A different method of holding artwork has been developed: the electrostatic copy holder. This has its own power source and consists of a sheet of translucent material which is given an electrical charge. The acetate cels are pressed together by moving a baton across them that imparts an opposite electrical charge. They will then cling together and eliminate all the shadows that can otherwise occur. Yet another method is the vacuum holder, although this is unsuitable for transparent material.

The camera

The registration camera used on rostrum consoles is similar to that used in the smaller optical printers, described in the section on duplicating. It will have a more elaborate viewfinder with a ground glass reticle showing a wider range of formats. The standard Forox 46/35 reticle has markings for superslides, 35 mm slides, KS slide mask cutoff, filmstrip, TV safe area and grid pattern. The reticles are interchangeable and for film animation a reticle showing 16 mm and Super 8 mm can be used.

Because of the variety of standards and formats in audiovisual the film movements also need to be interchangeable. For multivision 35 mm and 46 mm are the most commonly used and the camera must be easily adaptable for either of these. Fixed pin registration is provided on each of the aperture plates, giving an accuracy of plus or minus 0·005 mm (·0002 in). Because the pins do not move, the film has to be eased on to them by a pressure plate and removed by a stripper plate. In spite of this operation, speeds of up to 120 frames per minute for slides are possible. When the shutter is turned off for rewind, this advance rate is doubled.

The shutter can be either a fixed-size 180° rotary shutter which is synchronized to the film advance speed or it can have an additional dissolving shutter which will allow fades and dissolves to be made on movie film. If a slide show is to be transferred to film, this can be done only by reshooting the slides on a rostrum. In this case the camera will need to have an animation-type variable rotary shutter. This will be equipped with dissolve and logarithmic fade scales for 25 and 32 frames (1 and $1\frac{1}{3}$ second). If longer dissolves are needed, these can be used as a guide. The dissolving shutter is driven by the main rotary shutter and can be adjusted while the camera is running.

The clip-on magazines on top of the camera must be interchangeable to accept the different widths of film. They will also have a film notcher to mark the beginning and end of a run so that the exposed material can be located in the darkroom. A film cutter will also be provided that enables the

147

take-up magazine to be removed without unlacing the film.

The whole camera on some models can be rotated 90° to allow filmstrips to be made with the artwork facing the operator.

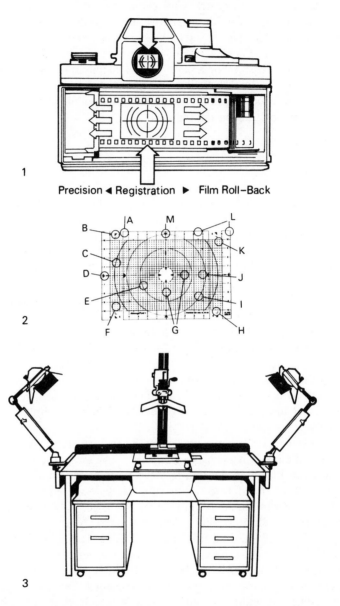

Registration photography using 35 mm cameras. 1, Nikon camera adapted to take pictures in register. 2, Slidemagic recticle showing information about slide formats, special slide masks, and audiovisual and TV live areas. 3, Camera mounted on rostrum.

Remote controls

The control module can be either a free-standing console which can be positioned to the right or left of the cameraman (useful for left-handed operators) or it can be a fixed panel attached to the table top. To this position all the controls for camera and lights are remoted, including a variable speed control for raising and lowering the camera head, together with an up/down switch. An exposure counter measures the number of frames exposed. When the counter reaches a preset number the camera automatically stops. The camera can be set to either the continuous run or single frame mode and there is a switch for forward or reverse operation.

Both the copy lights and the transparency illuminator can be switched to a bright or dim intensity level. At the 'dim' position they will idle at a reduced voltage until the exposure is about to be made when they snap on to full brilliance. Most control panels will give the option of strobe light being used and will include a sync input for this purpose.

Exposure

A photo cell fitted to the camera can give an exposure reading. However, the cameraman frequently makes tests to check that exposures are correct. Using a camera day after day with fixed lighting usually means that the cameraman can tell within a half stop how much exposure is needed. An exposure chart supplies the information for the different types of filmstock and standard top and bottom lighting intensities.

For quantity runs, a grey scale should be recorded at the beginning of the film with the exposure at the standard setting. This provides an immediate check on whether any density or colour balance faults are the result of poor exposures or incorrect processing.

The cameraman should keep a detailed record of everything that he shoots. A rostrum slate should be written out for each run of film. This will state the production name and number, date, filmstock, exposure details, field size, filter settings and light settings. A copy of this should be sent to the laboratory and a copy kept with the laboratory report.

Colour temperature control

Just as a grey scale is recorded on to film so a standard colour scale can be photographed. This shows the primary colours: red, green and blue, and the subtractive primaries: yellow, cyan and magenta. If a film is processed normally and a 'middle light' print is ordered the colour is set at 25-25-25. The grey scale should be neutral, without a colour cast. If there is colour in the grey scale when the print is made with a 25-25-25 setting then the colour reproduction is incorrect. For instance, it may need to be printed with a 29-23-23 setting. Since the average is still 25 the exposure must have been correct but the colour temperature of the lamps did not match the filmstock being used.

A colour temperature meter can be used to give quite an accurate reading although they are more suitable for other aspects of photography. These have a photo cell and either two or three filters to give a colour temperature reading in Kelvins is given. Tungsten halogen lamps produce a light at 3200 Kelvin but changes can occur with voltage fluctuations and as the lamp nears the end of its life. The lamps on the rostrum should be periodically checked for any changes in colour temperature.

The most reliable means of determining correct colour balance is the wedge test where exposures are made with progressive alterations to the filtration. This should be repeated using variations over about 2 f-stops since the degree of filtration affects the exposure. If you take out 50% of the red light with a cyan filter then overall you are reducing the light intensity by one-sixth. In the same way magenta will stop green light and yellow will block blue.

9   Slides

You need a schizophrenic attitude to slides. On the one hand they are images, the equivalent of words and sentences to a writer; on the other hand they are objects which have to be individually created and cared for. At different times during production you need either one attitude or the other.

The mounting, cleaning, numbering, duplication and storage of slides is scarcely the most exciting aspect of the medium. Its closest equivalent in everyday terms is housework, without which of course the design and style of the most wonderful home would be obscured. But just as there are domestic appliances in the home, so there are a number of machines and methods which can ease the work of 'slide care'.

35 mm registration slide mount.

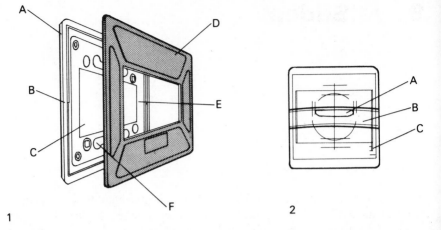

1, Standard 35 mm glass slide mount. A: White heat-resistant plastic faces projector lamp, prevents heat build-up; B: Channel snaps; C: Glass keeps slide flat and in focus for automatic projection; D: Grey slide faces lens to prevent internal reflections; E: Interchangeable metal masks; F: Centering guides to keep film in place.

2, 35 mm projector levelling slide. A: Air bubble; B: Oil; C: Format guide.

Mounting

During production slides will be frequently touched, sorted, scrutinized, even occasionally dropped, and it is essential that basic rules are observed. Firstly, individual transparencies must be protected. Both sides of a slide, celluloid and emulsion, are subject to scratches. The celluloid side will also attract dust like a magnet attracts iron filings. Any subsequent attempt to clean it can result in damage. For this and other reasons, slides should be mounted between glass.

Numbering

Once slides have been concealed in glass mounts they immediately become extremely difficult to find. As soon as they are mounted they should be numbered. In itself this is not as easy as it might as first seem. There are already too many numbers in the audiovisual business. The slide trays of each projector are numbered from 0 to 80. There are cue numbers in the script and storyboard. In a multivision show a slide change might involve a number of slides. To complicate the matter still further, different projectors may eventually carry different amounts of slides.

152

Slide handling can be the most time-consuming part of programme production. What is needed is a numbering system which will be helpful both before slides are loaded into trays and afterwards.

Two projector shows do not present a particular problem because apart from superimpositions, slide changes alternate from one projector to the other. It is the convention to label the projector that contains the first slide of the programme 'projector A' and the other one 'projector B'. Thus if the slides are numbered 1, 2, 3, etc. in order of appearance, then all the odd-numbered slides will be on projector A and the even-numbered ones in projector B. This is fine until you come to a superimposition. Then, for instance, projector B will show a slide at the same time as the other one is lit, and after the superimposition is over it will dim down and advance. From that point onwards all odd and even numbered slides will be in opposite projectors.

This can be very confusing to someone setting up a show who checks one of the slides, notes an odd number and places that magazine on the wrong projector. The magazines themselves should be numbered A and B, but over a period of time labels can easily come off or a magazine may be renewed and left unnumbered. The answer here is to always look at the first slides in the magazines.

A numbering system has also to take account of the fact that slides may be inserted at a later date. This flexibility that the medium allows may have been one of the reasons for using it in the first place.

Finally, the numbering system may be used as an indication of whether the slides are the right way up and the right way round. There are eight different ways of putting a slide into a projector and a number placed carelessly in the wrong position will only make the programme more difficult to load.

Bearing in mind all the above criteria, select the vitally important ones and number accordingly.

1. Slides should be numbered consecutively as they appear in the programme.
2. Numbers should be in the top right-hand corner as you view the slide on a light box.
3. Slide numbers must correspond to numbers on all written documents such as scripts, storyboards and cue sheets.
4. In single screen, two-projector shows this will be the cue number itself.
5. Inserts should be numbered, for example, 24A, 24B, 24C, etc.
6. Permanent inserts should be fully incorporated into the show and the remaining slides in the programme renumbered.
7. Do not try to indicate tray slot numbers that conflict with sequential numbers.
8. A and B projectors can be indicated by putting A or B on each slide in the bottom left-hand corner, upsidedown, as viewed on a light box.

More complex numbering is required for multiscreen shows. In practice, different producers have different systems of numbering, which are partly dictated by the format and the make of equipment they are using.

1. The additional information that must appear on each slide is the screen number.

2. Sequential slide numbers now relate to each screen.

However, when you have overlapping formats how are the screens defined? With the AVL Show Pro systems, the term 'screen' denotes a group of three projectors, regardless of the area which they cover. Therefore a single area with six projectors registering upon it would be called screens one and two.

The Electrosonic Microcue system shows the status of six projectors on each of 24 screens. The projectors on each screen are designated A, B, C, D, E and F. Since six is the normal maximum per screen area then it is possible to number the slides accordingly. Programming is now so versatile that slide numbering can scarcely keep pace with it. With individual projector control you can go from B to D, or from F to C at the touch of a button. Therefore a programme need not consist of equal numbers of slides in each projector tray.

Storage

Unfortunately, transparencies do not have great durability; the dyes that are used in colour films inevitably change in both hue and density over a period of time. Processing of films must be highly accurate in order to reduce the possibility of early deterioration. Particularly important are temperature, agitation, processing times, washing and the replenishment of chemicals.

Given that the laboratory processes the film to the optimum standards, how can the user best preserve his work? The main causes of deterioration are: *light*; *heat*; *humidity*; and *physical damage*.

Slides should be stored in the dark. Constant exposure to bright light will gradually fade the dyes, just as bright sunlight will fade a carpet. Keep projection times to a minimum. If your projection equipment has a rehearsal brilliance control, use it! This prevents damage during the time when slides are more likely to be left on-screen for periods longer than 1 minute. Have adequate numbers of slides for occasions such as lectures so that a single image is not retained on-screen for a long period. This applies particularly when using high intensity projectors. Also slides should not be left on illuminated light boxes for longer than necessary. Heat filters should never be removed from projectors and only lamps of the recommended wattage should be used.

In storing transparencies the temperature for short-term storage should not exceed 21°C (70°F). Very long-term storage in archives involves refrigeration of the material to −18 to −23°C (0 to −10°F) at controlled humidity. This prolongs their life by many years. A more thorough method is to make black-and-white separation negatives, one for each of the three component colours. When these are stored under ideal conditions, the silver density images will last unchanged for hundred of years.

Humidity can present one of the biggest problems in storage; both high and low humidity can be harmful. The recommended level of relative humidity is between 15 and 40% RH. Slides should not be stored in basements and attics unless these areas are specially protected against extremes in temperature and humidity.

Low humidities can be harmful because the film will become brittle and have a tendency to crack. High humidities promote fungus growth which

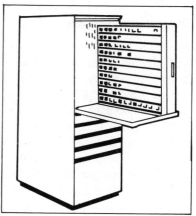

Slide handling and storage. 1, Slide sorting on an illuminated editing tray. Portable trays are useful since they can be stacked one above the other during production. Trays with horizontal supports allow use on vertical (as well as flat) light tables. 2, Storage. Small numbers of slides may be filed away in binders with transparent envelopes or clip-on slide holders. This method is convenient if a programme needs to be sent through the mail. 3, A professional storage cabinet with vertically mounted sheets each holding 120 slides. This has the advantage of instantly displaying this large number of slides against an illuminated background.

can irreparably damage the images. In tropical conditions extra precautions have to be taken.

Slides should be stored in moisture-proof airtight boxes, preferably with a metal lining and a tight-fitting lid. They need not be vacuum sealed but air should be allowed to circulate in the box. Silica gel may be added to absorb excessive moisture if the boxes are loaded in unavoidable humid conditions. The silica gel will need to be renewed at intervals, say, every few months, if it is to remain effective.

Apart from the expense involved, proper storage can do no harm, even in temperate countries. The expense is small in comparison with the cost of reshooting and in any case historical value is irreplaceable.

Storage envelopes

If exposed slide film is unmounted it should be placed in transparent sleeves. At high humidities the emulsion can develop shiny spots where it has been in contact with the sleeve. These can be removed by washing the film in water at 18–24°C (65–75°F) for a few minutes. The film is then given a final rinse in a flowing agent and dried.

The control of humidity during long-term storage is achieved by the use of specially produced storage envelopes. These are made from aluminium foil, paper and polyethylene which gives moisture and vapour protection. The film is brought to a 15–30% relative humidity at below 21°C (70°F) and sealed with a minimum amount of air.

In storing slides you must guard against unforeseen physical damage, particularly from chemicals. Keep slides well away from the darkroom and do not allow any developers, fixers or chemicals to be stored in the same room. Never use envelopes or other containers that are not specifically approved for photographic use. Most ordinary papers contain some harmful chemicals. Not only insects, but insect repellents can damage film. Dust, dirt, scratches, finger-marks and harmful vapours will also take their toll unless you guard against them.

Storage cabinets

Slides may be stored in their original slide trays, and it is probable that most programmes are stored in this way. This makes for quick retrieval and instant access without the need for reloading.

However, magazines are not inexpensive and many production companies find themselves with thousands of pounds (or dollars) worth of them in quite a short time.

Kodak Ektagraphic slide trays, being unenclosed, give inadequate protection against dust. A useful alternative is the reversible slide tray which has the advantages of being both cheap and dustproof. This type of tray is not used during projection, but solely for storage. The slide set may be transferred to it by capping the reversible tray on top of the projector magazine and emptying the contents into the slots provided.

The alternative to this method is flat storage. This is essential if the slides need to be individually viewed against a light source. The cheapest method is to load them into plastic envelopes which are then suspended in a metal filing cabinet.

A better method, especially for housing the master slide set, is to use a special slide storage cabinet that has been designed for the purpose. This allows hundreds of mounted slides to be stored vertically at a convenient position. Each set may then be pulled out and viewed against a light source.

If slides are stored in this way they need to be catalogued and numbered. Old programmes can be broken up and used as a basis for an extensive slide library. The numbering system used needs to be capable of expansion as the slide library grows. The recommended retrieval system is a decimal classification, like the Dewey system which is used in book libraries. To begin with, each major branch of interest if given a number. For example:

1. Art
2. Industry
3. Science

These are then sub-divided into more specific subject areas, for example:

2.0 Industry
2.1 Fuel
2.2 Manufacturing
2.3 Service Industries, and so on.

Again, these can in turn be sub-divided, as:

2.10 Fuel
2.11 Gases
2.12 Coal
2.13 Electricity
2.14 Oil

Each list will be in alphabetical order and retrieval of any particular slide is relatively easy. It is not necessary to write the full code on each slide. But it is essential to indicate the type of film used to take the shot, together with the date, and production number code. A card should be provided which lists all the different categories in the master set.

Each slide will therefore have the following information:

1. Film type
2. Date
3. Production number for year
4. Original number in programme

It will now be simple to find any duplicates which are kept in stock by seeking out further copies of the programme in the bulk storage file. For this purpose ordinary filing cabinets would be used. A note should be made of any duplicate which is taken from the bulk store and all master slides should be returned immediately after duplicates have been made.

Duplication

It is sometimes said that you should never project originals. However, it is often very convenient to shoot a number of original sets which may well be a sufficient quantity for the limited distribution of an audiovisual show. Always retain a master slide set, which is not projected. It is from this set that you can make further duplicates.

With care, a duplicate can almost match the photographic quality of an original. There are, however, a formidable list of inaccuracies which can prevent this quality from being achieved. Chief among these are: *contrast exaggeration*; *false colour balance*; *inaccurate exposure*; *lack of field flatness*; and *uneveness of illumination*.

A number of methods may be successfully used for duplicating slides. The choice of method really depends upon the quantity of work that has to be done. Starting with the least expensive and most time-consuming, these are the different methods that are used:
1. Photographing front-projected image.
2. Photographing rear-projected image.
3. Using extension tubes or bellows and slide holder.
4. Using a specially constructed light box.
5. Using an adjustable slide duplicator.
6. Using an optical printer.

Front projection

This immediately involves an initial compromise; the projector and camera lenses cannot share the same optical axis. They should be placed as close together as possible so that the keystone distortion is kept to a minimum. The slide should be projected on to a brilliant white screen a few feet away at right angles to the mean optical axis. If required, you can select parts of the projected image. The projector must be on full brilliance and a reflected light reading should be taken. The colour temperature of the projector lamp closely matches that required by colour tungsten film.

This method will require some trial and error. Be careful to note by how much the viewfinder on the camera crops the image. To reduce the problem of parallax choose a single lens reflex camera.

Rear projection

Throwing the image on to a rear projection screen eliminates the problem of keystone distortion but increases the difficulty of even illumination.

The camera is pointed at the matte side of the screen, which should be of a recommended rear-projection material. The projector behind the screen should have a focal length of 85 mm or longer. However, whatever lens is used there will inevitably be a concentration of light at the centre of the image. This is always so with rear projection but when seen by the human eye it is not so noticeable as when recorded by a camera. The camera does not have a brain to make any compensations.

Because of the hot-spot problem, this method is the least successful means of duplicating a slide and should only be used in the absence of other equipment.

Extension tubes

Like the previous two methods, slide copying with extension tubes or bellows fitted to the camera is strictly an amateur means of duplication. However, all the major manufacturers of professional 35 mm cameras make slide copying accessories and it is presumed that they are used.

The use of bellows, extension tubes or close-up lenses requires experiment with the particular equipment that you have. The first point to check is whether extension tubes will allow you to obtain a sharp image. The camera lens may not have been designed specifically for close-up work. Secondly, the aperture of the lens will need to be opened by at least two stops to compensate for the decreased angle of view.

No light should be allowed to reflect from the slide mount, or from the emulsion, into the barrel of the lens. A light source is required that will match the colour temperature of the film. With daylight film, electronic flash is suitable and this will tend to keep the contrast build-up to a minimum. The flash is best diffused by bouncing the light off a white reflector. Alternatively, a tungsten light source is easier to control for composition of the picture, focusing and measuring the exposure needed. A projector lamp, with a directional reflector behind it and opal glass to diffuse the light is satisfactory. In all cases, make tests and bracket your exposures.

Light box

One of the most practical ways of duplicating a slide is to use a normal copying stand and a light box. The light boxes which are used for slide viewing are of no use because they illuminate too large an area. It is simple for, say, an art school department or a home user to make a light box with either one or two tungsten-halogen lamps encased in a small housing with good ventilation. It is necessary for heat-resistant glass to be placed between the slide and the lamps to prevent damage to the film. The box should have a slide carrier on top which will perfectly cover the aperture of the opal glass area. The box should be white inside and matte black outside. Details of many such light boxes may be found in photographic guides and magazines.

Slide duplicator

This is the first professional method to be considered. Adjustable slide duplicators are made for use with single lens reflex cameras fitted with a bellows and an enlarging or reproduction lens. Their main features, common to most makes, are: two light sources, one for focusing and flash for exposure; an adjustable column for vertical mounting of the camera; registering slide carrier with a range of masks for 35 mm, filmstrip, etc.; and a contrast control device.

A slide duplicator is a compact and useful tool but in essence is simply an accurate solution to the problems encountered in the methods described above. It will enable you to make duplicates more quickly and with a decreased failure rate.

The biggest advance has been in contrast control, for which different manufacturers have alternative solutions. An ingenious method is the use of a fibre optic to give controlled fogging to the film. A fibre optic conveniently carries light direct from the control box to the bellows of the camera where a small diffuser spreads the light evenly across the film plane. The amount of light which is carried can be varied by an adjustable control, and is achieved by a built-in neutral density filter ring on the base unit.

An alternative method of giving selective fogging to the film is to use a separate light source projecting on to a deflector set at an angle of 45° to the main light path. The second light source can be a low power flash tube which operates in sync with the main light source. The main practical advantage of this system is that it requires no modification to the lens barrel or bellows.

With a colour slide duplicator you can correct both colour balance, by using filters, and density by compensating with the exposure. In this way the duplicate can actually be a better quality slide than the original. You can also enlarge a section of the image, make copies of composite slides, superimpose title slides, add colour to black-and-white or make colour internegatives. The duplicator should be used as a creative tool and there is no limit to the visual effects that you can achieve. Several great photographers have used the duplicator to produce a final picture from which a colour print is made. All of this is made possible by the accuracy of the equipment and the labour-saving devices of exposure, contrast and colour temperature control.

Sophisticated duplicators make the process of straight duplication extremely simple. The slide is placed on the light box in the holder, the tungsten lamps are switched on and the lens is accurately focused by turning a wheel adjustment. On some systems a ratio scale can be fitted to the column. This indicates the degree of magnification that will be obtained from the particular focal length of lens in use.

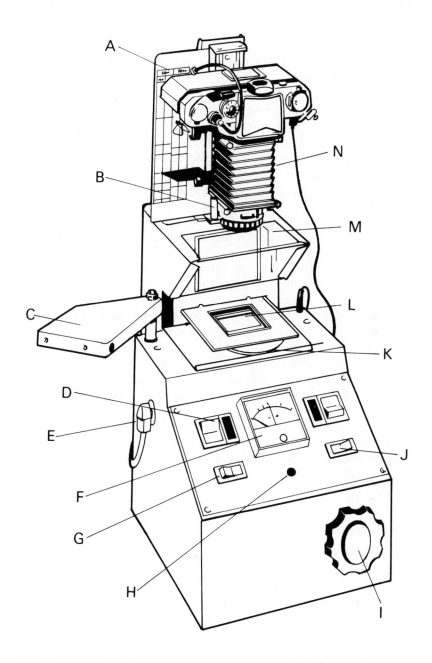

Slide duplication. Small desk-top slide duplicator. A: Magnification/exposure scale; B: 2-way movement available on bellows track; C: Photo-cell; D: Power on/off switch; E: Power supply socket for contrast control; F: Exposure indicator; G: High/normal flash intensity switch; H: Trimmer control; I: Light control knob (for exposure control); J: Open flash switch; K: Filter drawer; L: Universal frame with interchangeable slide holders; M: Contrast control unit; N: Bellows (with camera and lens mounting rings). Interchangeable to adapt all suitable 35 mm cameras and lenses.

161

Exposure

A built-in exposure meter gives an accurate measurement of the exposure required. The strength of the flash is determined by measuring the amount of the tungsten focusing light that is transmitted through the transparency. For this purpose a photocell is swung across the top of the slide. Be careful that it does not overcompensate for slides which are meant to be very dark or very light. The flash setting is adjusted by centering the needle of a meter dial by means of a control knob. The photocell is swung back, the focusing light is extinguished and the exposure made.

The use of a variable-power electronic flash enables the operator to work at a fixed aperture. If the exposure needs to be varied, this can be done by reducing or increasing the power of the flash, by far a preferable method to changing the f-stop where there is a danger that the lens will be moved out of position.

Electronic flash also has advantages in that the heat problem encountered in building a home-made light box illuminated by tungsten halogen light, is eliminated. The slides will not buckle which is the cause of them drastically going out of focus. Electronic flash enables daylight colour films to be used which are balanced for a colour temperature of 5 600K. This does not vary appreciably with the strength of the flash.

Finally, a word about the limitations of the duplicator. For very long runs the film capacity of an ordinary 35 mm camera is insufficient. Constant rewinding and reloading is tedious as is manually winding on the film for each shot. One other drawback, particularly applicable to audiovisual users, is the absence of exact registration. With more effects being used that involve precision register the compact slide duplicator is no longer adequate for all the work of a studio. However, the price jump between this and an optical printer is considerable and has to be justified by the amount of work to be done.

Optical printer

The 35 mm reflex camera is by tradition and design the tool of a photographer working on location. Its incorporation into post-production work is a big compromise. Even the most precise 35 mm cameras are produced in great quantities and marketed at relatively low prices. Relative, that is, to equipment that has been designed for specialist applications. The optical printer is a good example.

Basically, an optical printer does exactly the same job as the other equipment described above. But just as you cannot necessarily deduce from the appearance of a traveller whether he has arrived by bicycle or Concorde, so you cannot necessarily tell whether a slide has been duplicated by a home-made light box (costing the equivalent of a few packets of cigarettes) or on an optical printer (costing the equivalent of a small house). However,

20 000 miles and 20 000 slides later, both the traveller and the photographer will be able to tell you the difference!

From the point of view of cost, the optical printer does not have the advantage of using ready-made and relatively inexpensive components. Its camera and light source are conceived as a complete concept, designed to do a specific job: to copy accurately and quickly. Most optical printers will be versatile systems with a whole range of accessories that enable a photographer to copy most types of visual material, not merely 35 mm transparencies. There are two main types of printer: desk-top models and floor standing consoles.

Desk top models

The smaller type of optical printer consists of a base unit, containing a quartz light source and a control panel, from which extends a camera support column about 1 m (3 ft) high. The camera is fitted with magazines capable of handling at least 30 m (100 ft) of film. Both the supply and take-up units are removable which allows the operator to load fresh material easily. The film enters the camera through light-trapped velvet rollers and takes a path around tensioning rollers, through the exposure gate and out into the second magazine.

The transport mechanism allows high speeds. Around 40 to 60 frames per minute is average. The camera can also step a half-frame at a time for some filmstrip applications. On sophisticated systems there is 'no advance' button which enables multiple exposures to be made. You can remove the aperture plate to accommodate different sized apertures for alternative formats.

The accuracy of the camera depends on whether it has fixed pin registration. This is normally a feature of the console models which are large enough to accept an animation table. However, there is no reason why a desk top model should not provide really accurate film registration for precise optical work. This is an important feature to observe in the manufacturer's specification. Film advance accuracy will be about plus or minus ·025 mm (·001 in) without fixed pin registration; with it, the accuracy is increased to plus or minus ·005 mm (·0002 in).

This degree of precision can never be matched by a normal 35 mm camera. An ordinary camera relies upon the sprocket wheel to position the film. If you examine these sprocket teeth closely you will see that they are smaller than the perforations of the film. Since the tooth of a sprocket wheel enters the perforation at an oblique angle this decreased dimension is necessary. A 35 mm camera is also designed to take account of shrinkage that occurs in photographic film. Therefore, no 35 mm camera, unless it has been heavily adapted, can provide the degree of accuracy that is given by an optical camera.

The precision of the film registration is of course quite redundant unless the material to be copied is also accurately placed. A slide carrier positions successive slides above the light source and an equally precise viewing system is attached to the camera for line-up purposes. Most viewers use a reflex mirror system which shows what is seen by the lens and transmitted on to film. The image is shown on a ground glass reticle (sometimes called a grati-

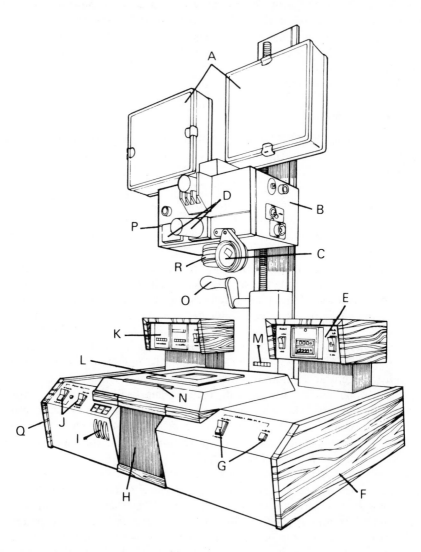

Slide duplication. Desk-top optical printer. A: 400 ft magazines—supply and take-up; B: Sickles camera; C: 4× magnifier; D: Photometer and light meter package (optional); E: Power/illumination module; F: Swing-down service panel (right side); G: Exposure controls; H: Light mixing chamber access door; I: Diallable filtration wheels; J: Film advance controls; K: Counter module; L: 4×5 in colourhead; M: Camera position indicator; N: 5×5 in filter drawer; O: Positive adjustment handle; P: Film notcher; Q: Swing-down service panel (left side); R: 55 mm f3·5 Micro-Nikkor lens.

cule) which is etched with a grid pattern and with the different frame formats that are likely to be needed. The reticle may also indicate the TV safe action and title area and the field sizes that correspond to the height of the camera and hence the size of the area being photographed.

The viewfinder is at eye-level for normal-sized copying and has a magnifier to help the operator see the image clearly for focusing and composition. In operation, alignment can be extremely accurate. On the larger models where the column is around 3 m (10 ft) high and eye-level viewing is difficult, the reticle grid can actually be projected on to the material to be copied.

Height adjustment is normally by a crank handle. The mechanism should be balanced with a good gearing system that prevents backlash and camera drift. An indicator tells you the exact height of the camera so that particular positions can be quickly repeated.

An optical printer has full facilities for colour filtration. The most convenient method is a colour head with which you can control the degree of magenta, cyan and yellow filtration simply by setting numbers on a dial. A filter drawer may also be included for additional filters and infra-red and ultra-violet wavelengths will be filtered at the light source.

A light meter built in to the camera calculates exposure. Normally, manual setting of the exposure is necessary although this operation may be remoted to the control panel on the base unit. Since the camera is equipped with a rotary-type shutter, as used in movie cameras, the exposure is determined by setting the iris of the lens.

There are many accessories which can be used with the basic optical printer. Some models accept alternative magazines to hold 16 mm or 46 mm film. Choose one to fulfil your present and future needs. Copy lights may be added for the photographing of opaque artwork. And there will be a full range of facilities for producing filmstrips, such as a loop transport system that will advance a filmstrip master for release printing. Other attachments include an automatic slide advance for holding the slides in correct order, ready for duplication.

The technology of printing has advanced to a degree which makes the whole operation as quick and simple as taking a Polaroid photograph. It is carried even further with the top-of-the-range models that are favoured by audio visual professionals.

10   Sound recording

By far the majority of audiovisual programmes have pre-recorded sound on one or more tracks of magnetic tape. Live speech and music have been used, as have other sources of pre-recorded sound, such as gramophone records and optical soundtracks. However, the nature of the audiovisual medium, with its multiplicity of programmes made for specific audiences, has employed the convenience of magnetic tape recording for both production and playback.

Tape base

Magnetic audiotape consists of a coating of powdered metallic oxide on a *base* of polyester plastic. In recording, the metallic oxide particles are magnetized by an electronic signal and thus hold audio frequency and relative sound intensity information. The electrical signals also contain a bias signal which is constant although greater in amplitude than the audio signal.

Magnetic tape base used to be acetate, which was more subject to breakages than the currently used polyester plastic. The difference between the two materials may be noted by the fact that while acetate tended to break, polyester tends to stretch. This is not a serious problem because tape transport mechanisms are designed to handle the material with care. However, the strength of magnetic tape is directly related to its thickness.

Long-playing tape is thinner than standard tape. The reason for a variety of thicknesses is to enable different lengths of tape to be wound on standard sizes of take-up spools. The average thickness is around 1 mil (·025 mm, ·001 in) and magnetic tape may be as thick as 1·5 mil. If very thin tape is used there is not only a danger of stretching it, but also of 'print through' where the signals are transferred through the tape on a tightly wound reel. Professional quality 1·5 mil tape should be used whenever possible.

Tape coatings

Magnetic tape coatings are manufactured in a number of variations. The most commonly used are the following:

Ferric oxide (iron oxide). The most widely used in recent years. All professional tapes are now high quality, low noise ferric oxide coated.

Chromium dioxide tape. Widely used in cassettes and in video recording because of the need to cover a far wider frequency range with less material.

Lubricated tape. For cartridge use.

The coating, in each case, consists of tiny magnetic particles which are dispersed within a binder. To every inch of cassette tape there are about 20 thousand million particles. Each of these is a magnet with its own North and South poles. These polarities can be switched by magnetic fields from the record heads, and in turn, they create a magnetic field at the surface of the tape which is read by the playback head. The particles must be of small and uniform size for high resolution of the magnetic fields.

A blank tape, that is, one with no recording on it, has the particles in its coating arranged in a completely random way. All that can be played back is tape noise or 'tape hiss'.

The properties of the particle coating are analysed under three main headings: *saturation level (Bs)*; *remanence (Br)*; and *coercivity (Hc)*. *Saturation level* is the point at which no further magnetization is possible because all the particles have been aligned in the same direction.

Remanence is the measurement of the degree of magnetization remaining after the tape has passed the record head. High remanence is desirable because it means a strong playback signal, low hiss and low distortion.

Coercivity is the resistance to magnetizing or demagnetizing. It is a measure of the force needed to remove the remanence of the coating.

These properties are shown by applying a magnetic field which varies between high positive and negative levels. The resulting *hysteresis loop* shows a graphic representation of the characteristics of the coating.

Track width

With photographic emulsions the quality or definition of the picture is related not only to particle size but also to the actual size of a negative or transparency. The same is true of magnetic tape where track width and tape speed both affect the quality. Track width is governed in turn by two factors: the overall width of the tape and the recording/playback head format:

Tape can be 2, 1, $\frac{1}{2}$, $\frac{1}{4}$ or $\frac{1}{8}$ in width. At the top of the range 2 in is used for broadcast quality video recordings and at the lower end $\frac{1}{8}$ in is the width of a normal audio cassette. Certainly, the difficulty of handling a very wide tape involves enormous expense in designing powerful, yet precise, transport mechanisms which would not be within the reach of most users of magnetic tape. It is worth noting that a good quality $\frac{1}{8}$ in cassette when played on one of the better sound systems gives far better sound reproduction than you are likely to hear on a television set, even though the TV studio has used the

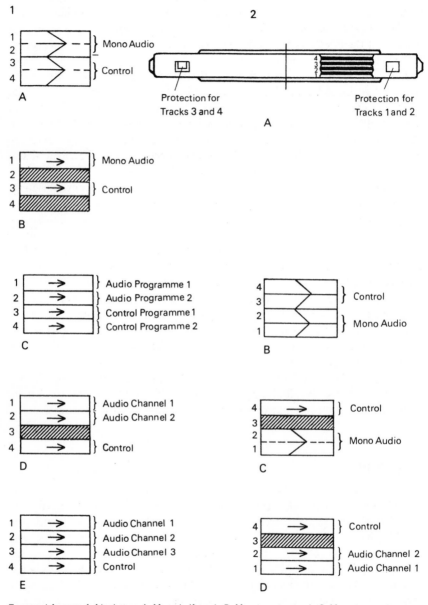

Tape track formats. 1, $\frac{1}{4}$ inch tape. A: Mono half-track; B: Mono quarter-track; C: Mono two-programme; D: Stereo; E: Triphonic. 2, $\frac{1}{8}$ inch cassette (A). B: Mono half-track format; C: Mono audiovisual format; D: Stereo.

highest quality recording facilities available. Never forget that it is subjective quality which is the most important.

In the audiovisual industry, $\frac{1}{4}$ and $\frac{1}{8}$ in are most widely used for playback. It is only when multitrack sound is required that wider tape becomes necessary. In considering track formats remember that special conditions apply with audiovisual use. A control track is required. It is vital that this control track is played separately from the audio tracks. The ideal is no 'cross-talk', that is the picking up of signals from another track. Control signals are within the audible frequency range and will naturally be heard if there is any cross-talk between tracks. The second main consideration is that, unlike the uses of magnetic tape in, say, home sound reproduction, only a single pass of the tape in replay is necessary. It is unlikely that any producer, amateur or professional, would want to have one programme in one direction of the tape and a second programme in the other direction. In home recording tape is a major expense. In audiovisual the highest quality tape is only a fraction of the budget.

Track configurations have been developed to take account of these criteria.

Tape speed

The other factor to consider if you want to obtain the best quality of recording is tape speed. Here the standards are as follows:

1.	15 inches per second	*studio reel-to-reel*
2.	$7\frac{1}{2}$ ips	*NAB cartridge and reel-to-reel*
3.	$3\frac{3}{4}$ ips	
4.	$1\frac{7}{8}$ ips	*cassettes*

For music reproduction it is best to record the original on the fastest speed possible. Master recordings should always be made on $7\frac{1}{2}$ ips at least. The reason for this is that although the slow running cassette may give adequate quality during reproduction there will inevitably be loss of quality in making transfers from one type to another. If there is, say, a 5% loss during transfer then this is eliminated by using a higher speed for the original. Most audiovisual shows need to be duplicated. Even the one-off show ought to have a spare tape (or two) for emergencies. Packaged systems that include a cassette deck on which you both record and play back are intended for only very simple one-off shows.

Splicing

Reel-to-reel tape is used therefore for making the master recording. The faster the speed of the tape, the easier will be the task of editing it. During editing, the tape is physically cut and joined together and it is often necessary to be very accurate in choosing where to make a cut. The exact point is marked with a wax pencil. The sections of tape are immediately joined up and wound on to a spool. It is not practical to handle long strips of unwound tape in the manner of movie film.

Tape should be cut on a metal splicing block using a sharp razor blade (preferably a one-sided blade!), and joins are made with $\frac{7}{32}$ in splicing tape.

Locating the actual point at which a cut should be made involves a little practice before the operation can be done quickly and skilfully. The secret is to use headphones, and the pause control on the tape recorder, and to turn the reels manually to find the exact place. When you start it is hard at first to distinguish speech when advancing the tape by hand. But eventually it is easy to find a particular word and to edit to within a split second. Some tape recorders are not intended to be used for editing and it is not possible to 'inch' the tape by hand nor to monitor sound on headphones while the deck is in the 'pause' position. Check that the tape recorder you intend to use is suitable for the job.

Tape recorders

Whatever make of tape recorder is used in audiovisual production it will have a number of features which are common to all machines. Like a slide projector a tape recorder is both *mechanical* and *electronic*. Its mechanical elements consist of a transport system that passes the tape across the record and playback heads. The controlling element in the system is the *capstan*, a small motor driven spindle that rotates at an exactly regulated speed. The tape, loaded on the supply reel, on the left-hand side of the deck, passes between the capstan and a *pinch-wheel*, which pulls the tape at an accurate speed across the heads. The tape is then rewound on to a take-up reel on the right-hand side of the deck.

When the tape is wound on in the fast-forward mode or reversed in fast-rewind, it is held away from the recording and playback heads. There are a variety of heads used in tape recording. Some machines have three separate heads: erase, record and playback. Although these recorders will function more efficiently you have to remember that there is a gap between the heads and that you need, for instance, to offset control pulses by a fixed amount. Other recorders have the functions of record and playback in a single head.

Whether the head has one, two or three functions, the only part of it which is active is the extremely fine gap at its centre. The variations in the electronic signal across the head cause the tape to be magnetized. During playback the reverse process takes place, that is, the magnetic field is translated back into electronic signals. These are amplified and sent to loudspeakers which turn them into soundwaves.

The head is the most delicate and precise part of a tape recorder. It must be accurately aligned, a job which should be done only by an expert, and regularly cleaned. It will also need to be demagnetized by passing a demagnetizer close to it. Cleaning should be according to the manufacturer's instructions. A cleaning fluid, derived from alcohol, is wiped across the head with a cotton pad. Dirty heads are a major source of problems in audiovisual control, especially in playback. Brief duration pulses can be missed if an accumulation of coating has been allowed to build up on the head.

Recording meters

Common to all good tape recorders are VU meters that measure 'volume units' for each channel. A stereo tape recorder has two VU meters each having a white (minus) scale and a red (plus) scale. The 100% mark is where these two scales meet, and from which you can measure increases or decreases in volume. During recording the needles fluctuate over a segment of the scales. The associated record level controls should be adjusted so that the needles only occasionally venture into the red (overrecorded) segment.

When recording control pulses you should follow the manufacturer's instructions about recording levels. As a general rule, overrecording should *always* be avoided for control tracks. The normal setting is $-6dB$. If you need to amend a previously recorded tape the level must be adjusted to match the level on the tape. For this reason all master tapes should have a few seconds of set level of the control tone at the beginning of the tape. The new recording should be within 3dB of the original.

When using a VU meter, remember that the reading given is in proportion to the average energy content of the signal. It does *not* indicate peaks. This is the function of a pp meter which is found on some expensive equipment. For this reason there is often a danger of overrecording when using a *R*U meter and it is essential to record several seconds of set level to select the volume for control tracks.

One of the most frequent jobs in the audiovisual studio is the making of copies. This is known as 'dubbing' which is one of those jargon words that is (confusingly) used to identify several distinct processes. (It is a corruption of 'duping'.) Re-recording (transferring, dubbing, duping or copying) may occur during production of the tape with electronic editing, or at the end of production with the making of duplicate tapes, cassettes or cartridges.

Cartridges

Cartridges are used for completely automatic audiovisual installations in exhibitions, visitors' centres, or anywhere that unattended operation is required. They consist of a continuous loop of tape which has been cut to length. For reliable operation, lubricated tape is used. This has a graphite molyodenium disulphide or other coating enabling the tape to be wound in the cramped space of a cartridge.

The tape recorder which is used for loading the tape on to a cartridge spool needs to have a very gentle winding action. The tape should never be tightly

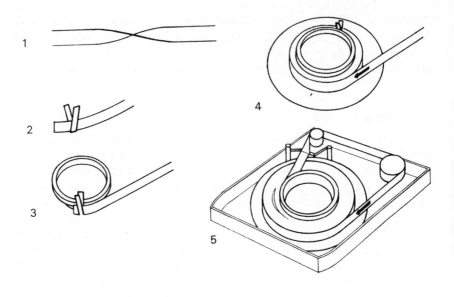

Cartridge loading. 1–5, Tape should be wound coating-side-out on to the platform of the cartridge. A half twist of the tape is made.

wound on to the spool. It is a good idea to use a recorder other than the one on which the recording itself was made. Some studios reserve an old machine for this purpose.

Cartridges are made to different standards and they may be intended for replay at $3\frac{3}{4}$ ips (mini-8 cartridge) or $7\frac{1}{2}$ ips. After the dub has been made you should locate the beginning of the recording and cut the tape 4 sec before the first (engineering) pulse or before the start of the soundtrack (whichever is the earliest). At $3\frac{3}{4}$ ips this will be 15 in (38 cm). At $7\frac{1}{2}$ ips it will be 30 in (76 cm). Wind on the tape until the end of the recording and locate the reset signal. This is the last piece of information on the tape. (With continuous tone dissolve units the reset signal is an absence of tone.) Cut the tape 4 sec after this point. Transfer the reel to the right-hand hub of the tape recorder (or to the right-hand hub of the special recorder used for loading cartridges).

Put the cartridge platform on the supply (left-hand) hub of the recorder so that the end of the tape can be attached to it. The tape must be wound coating-side-out on to the platform; to achieve this half twist the tape.

Attach the end with a V-shaped card which will later enable you to pull it out for joining it to the beginning of the tape and completing the continuous loop. When all the tape is wound on, connect the two ends with splicing tape.

A degree of slack should be left within the cartridge for reliable operation. In fact, after running a few times it will develop a noticeable 'eye' of slack in the reel. The amount of slack can be adjusted by cutting at the joint and

172

either pulling out more tape or winding it tighter on the spool. The lid should be fitted after running the tape several times to check that the winding is correct.

Cartridges have a maximum and a minimum playing time. Sufficient tape must be on a cartridge for it to function properrly. This may involve repeating a short show two or more times in order to meet the minimum duration requirement (normally around 6 min).

Microphones

The microphone is probably the most important single item in a sound studio. It converts the energy of sound into electrical energy by using one of several kinds of vibrating elements. The electric signal is then amplified, equalized, perhaps combined with other sound sources in a mixer, and recorded. The main types of microphone in general use are: *crystal*; *ribbon*; *moving-coil*; *carbon*; and *capacitor*.

Type of microphone	Principles	Directional response	Impe-dance	Use
Crystal	Crystal bimorph. Two crystal slides cut and placed together so that a piezoelectric voltage is created when the crystals bend in response to sound pressure waves.	Omnidirectional	High	In conjunction with low-cost cassette recorders
Ribbon	Narrow ribbon foil suspended in magnetic field. Output voltage is generated by movement of foil with air pressure *or* pressure gradient.	Bidirectional	High	Less popular today with improvements in other types of micro-phone. Good for voice reproduction, and location work in windy conditions.
Moving coil	Coil attached to diaphragm moves over fixed centre pole in field of a magnet.	Cardioid or omnidirectional	High or low	Most applications, particularly location work and hand-held work.
Carbon	Button of granular carbon in which resistance is changed by pressure on the diaphragm. A polarizing voltage is applied to the carbon to pick up the fluctuations in electrical resistance.	Omnidirectional	High	As with crystal microphones: not for high quality work.
Capacitor	Light flexible membrane diaphragm and a rigid backplate interact to produce fluctuations in capacitance and so generate an electrical signal.	Cardioid or omnidirectional	High or low	Whole range of types available for all uses, including miniature lapel microphones.

Four factors are particularly important in the performance of a microphone *directional response*; *distortion and frequency response*; *durability*; and *size, appearance and cost*.

Directional response

The directional response, or polar recording pattern, is the pattern of sensitivity. It can be plotted as a polar diagram in which the sensitivity of the microphone is plotted against the angle from the axis. A diagram shows the angle of acceptance for incoming sounds. There are three main types of pattern:
Omnidirectional. Response to sound from all directions. Measures air pressure.
Bidirectional. Response to sound from two sides only. Measures pressure gradient.
Cardioid. Heart-shaped response with broad field on live side. Combination of both pressure and pressure gradient.

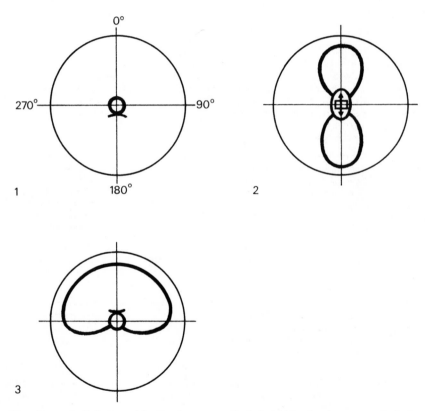

Microphones. 1, Perfect omnidirectional response. 2, Perfect bidirectional response. 3, Perfect cardioid response.

By combining the response pattern with sound baffles or reflectors, specialist microphones can have very exact directional response.

Distortion and frequency response

Ideally, a microphone should pick up all the frequencies present in the sound which is to be recorded. The electrical signal which it produces should be well above its own electrical noise level. The sensitivity of a microphone is the voltage per unit of sound pressure. If it is overloaded the microphone will distort the sound, although some types such as the moving coil design, can cope with overloading reasonably well. During distortion the electrical waveform ceases to follow the sound waveform.

Frequency response depends on a number of factors including:

1. The physical shape of the microphone, high frequencies are shielded by the body of the microphone.

2. Reflection inside the diaphragm housing, high frequencies can be distorted.

3. Resonances of air cavities between casing and diaphragm.

4. Cancellation of frequencies when the wavelength corresponds to the size of the diaphragm.

5. *Standing waves* formed in front of the diaphragm, and the resonance of the diaphragm itself.

So, the smaller microphones can be inherently more accurate than larger ones. In windy conditions any microphone needs a wind shield. Ribbon microphones are least sensitive to wind noise.

Durability

Durability is essential for uses outside the studio. You should check whether the types of microphone you are using are robust or fragile if you use them for location work. Moving coil microphones are particularly strong and will stand up to quite rough treatment.

Size, appearance and cost

Size and appearance are relatively unimportant for audiovisual uses unless the microphones are also to be used in video production or during a presentation. Cost, on the other hand, may be an important factor. The cheapest microphone is of the crystal type, which is not really adequate for professional use. The best value may be a good moving coil microphone while the best quality is obtained from a capacitor microphone (which is expensive).

Handling microphones

Microphones may be designed for particular handling techniques:

175

Lavalier type. For weaving around the neck on a cord, (in the style of Madame de la Vallière, a mistress of Louis XIV who wore her jewellery that way).

Floor stand or desk-top stand mounting.

Hand-held.

Boom-mounted, controlled by an operator. TV and film usage only.

Lanyard or lavalier microphones are popular among presenters because they enable the user to move about without having to be followed by a boom operator. The disadvantage is that clothing noise is easily picked up causing bumps and crackles in the sound. Weaving them requires some practice. An alternative is the extreme miniature lapel microphone which has a remote electronic section that can fit into a pocket. These are of the capacitor type and correspondingly expensive.

More freedom of movement still is provided by radio microphones that transmit the signal via a radio link. In this case the user has to wear a transmitter, aerial and power pack. The effectiveness of these microphones is subject to interference from other radio sources, particularly if talkback equipment is also being used, but it does do away with the cumbersome cable link to the sound mixer.

Hand-held microphones need to be omnidirectional in response because the working distance will vary. The moving-coil type is generally used and is made to be relatively insensitive to handling noise. The normal size is about 17·5 cm (5 in) long and 2 cm ($\frac{3}{4}$ in) diameter. It is responsive to a wide volume range although it is limited in sensitivity. More sensitive capacitor microphones are now available for hand-held work.

Purchasing microphones

The main recommendation must be: buy the best that you can afford. The microphone is the initial component in the recording/playback chain and it needs to be of high quality. Try out a variety of microphones, if possible, before making a purchase, and listen to the results. Do not try to use just one microphone for all uses.

11   Sponsor's guide

If you sponsor audiovisual shows you know that apprehensive feeling as you shake hands on the deal and go away to wait for the producer to make the next move. It can be like going on a long holiday, leaving your house to be redesigned and decorated in your absence—you might not like it.

The producer's role

Your company, your business is your creation. Even if you are a manager in a large corporation you are more closely involved than any outsider. An audiovisual producer, too, is a businessman. Of course, he will try to understand your point of view, but the satisfaction that he derives from his business is in creating superb multivision shows that will be admired for their own merit as examples of the medium.

At the heart of all business is the concept of exchange. The audiovisual producer must be allowed to achieve more than merely a reasonable profit for his work. He must be allowed to demonstrate his own particular style. Indeed, that is the best reason for commissioning him. His style, if it is good and appropriate, will enhance what you are trying to do. It will make a positive contribution that would be lacking if the producer was a 'yes-man' in your own payroll.

Multivision shows should be inventive, stylish, sophisticated, witty and entertaining. All of those qualities, you will note, have a degree of independence about them. They are not directly related to selling motor cars or informing people about the merits of natural gas. Many sponsors are willing to sacrifice the lot of them. They cut out all the invention, style, sophistication, wit and entertainment merely to cram in another dozen reasons why

Joe Public should spend his money on another freezer. But they are so mistaken, because Joe Public himself wants to be all of those things. He expects the products he buys to be all of those things and he is also independent and self-willed. Therefore, the independant stance of the programme, that unfamiliar voice you hear talking about *your* products, is already bridging the gap between you and your audience.

A potential sponsor took me to one side recently and confidentially asked my advice about finding a production company. 'I don't really want one of those fey young men who will tell me that it all ought to be done in blue.' I assured him that if there were any producers like that they would not have progressed very far in the industry. I did not tell him that *all* good producers are deeply interested in the aesthetics of the show. They just pretend from time to time to be interested in motor cars, ball bearings or health food products.

The working relationship with your producer is important, but before it can be established you first have to choose a production company. Very few sponsors approach a media company and say, 'I want a big multivision show'. They are more likely to want help in weighing up the pros and cons of the various media. They will approach a number of companies who make not only multivision shows, but films and video programmes as well. The sponsor presents his particular communication problem and the producer advises the best medium or combination of media to use.

This approach implies a certain degree of ignorance on the part of the sponsor. From the start he is allowing himself to be persuaded by a media man about the most fundamental part of the project. However, if the people he asks are very professional, and equally capable of producing brilliant video programmes, cine films and multivision shows, then the chances are that all will be fine. Unfortunately, many different skills are required in each of the media and it is unlikely that you have stumbled upon a protean genius who can turn his hand to anything. Most producers try to guide you towards the medium in which they are most at home.

The visual media

What, then are the unique characteristics of the different media? Leaving aside for one moment the apparatus required to play the different media we should note that film, video and audiovisual are all at different stages of development. Sponsored films have been made, effectively, for around 50 years. Video is much more recent. Although television for a mass market has been with us for a long time and has been intensively developed, video programmes for specialized audiences have been made only in the last 20 years. However, they have had the benefit of TV experience behind them. Audiovisual is the newest medium with a history of a mere 15 years experimental development, although the very first attempts were made far earlier.

Film

Film has the unique ability to condense the time in which an activity is seen to happen. The most important, relevant and exciting parts of that activity can be chosen for showing to an audience. This gives the opportunity of creating an almost totally believable drama, lengthening and heightening moments which in real life pass very quickly indeed. For this reason film is very rarely 'truthful'. Take for instance a motor race. A film of the subject may be made by, say, a tyre manufacturer. There are dozens of techniques which may be used in order to heighten the drama. A champion driver may have pre-recorded some off-the-cuff comments while relaxing in the bar. These can be used in the soundtrack while the picture shows him in the race battling against a wet, slippery road. How simple it will seem! All you have to do is to buy that brand of tyre and you can be just as relaxed under these conditions. Then, again, the champion always seems to be overtaking some-one. Other drivers spin off the road—in slow motion. They spend half the race, it seems, spinning off into the bales of straw. The champion's tyres are changed in the pits in 10 seconds flat. We don't see that the old tyres are badly worn; but we do see a fine close up of the new tread as it bites into the road and disappears into the distance. Our viewpoint is omnipotent, giving us the impression that we see anything and everything. Of course, we only see what we are shown, but the sheer delight in watching a film does not leave us a chance to reflect upon the *time* that has been left out.

Film, then, gives the illusion that time is being faithfully observed because we see something happening before our eyes. But in reality an editor and a director have selected, and maybe even contrived, in order to create a different reality. The director chooses the interesting and 'relevant' things that the participants have to say. If they are responding to a particular question, even that question may be cut out of the finished film. And it all could have been carefully scripted beforehand. It may all have been rehearsed and yet it will appear to be spontaneous. Everything that is said and that happens in the film will be relevant to the film's purpose. It is therefore essential for not only the producer and his team, but also the sponsor, to know exactly what the purpose of the film is supposed to be.

Video

Video may eventually equal film in the above characteristics. By video, I mean here, the recording of moving images direct onto tape from an electronic camera. However, and here is where the nature of the equipment creeps in, it is a more immediate medium. Anyone who makes a video recording is aware that he can have a playback of it simply by rewinding the tape. There is the implication of: 'I am talking to you, now.' This is unique to the medium and for that reason it is most often used as a 'talking medium'. The company chairman can sit down and speak to each person individually at the same time, and without the risk of everyone immediately asking questions back. A teacher can give a lecture or a scientist explain a process. It is likely that they will have other visual aids—film inserts, outside video recordings which have been edited, even slides. And video itself is the equal of film in showing an activity, providing that that activity is more confined in both time and space. In practical terms, video still lacks the atmospheric charm of film, This is because video production is given all the work which is required quickly. Until recently, far more expensive equipment was required for recording and the user overcame this by increasing the number of programmes made.

Multivision

Multivision, or audiovisual, stands apart from these other media. The term 'audiovisual' in this book always refers to a programme consisting of a series of projected slides accompanied by a sound track.) Multivision is not a poor man's film. It is rather a rich man's talking picture book. It can have a soundtrack very similar to that of a film which can either explain the visual images or provide a counterpoint to them. But because there is not the distraction of watching a recognizable activity, of seeing a real or contrived reality, there is far greater scope for making a more direct impact. The multivision producer is freed from the constraint of having to show (or conceal) the real world and the activities that take place in it. This means that he has all the advantages of a poet or novelist or an advertising copy-writer. He can stimulate the imagination of the audience. He can make them search their own experience and draw from them a direct and heartfelt emotional response. This does not happen with the majority of multivision, but I have noticed it on many occasions and it would seem to me to be an extremely important phenomenon. It can release what Maslow or Colin Wilson (or Leslie Buckland) would call a 'peak experience'.

It is extraordinary that multivision has the capability to provoke such responses, especially considering that most multivisions have a direct commercial connection. The themes which generally move us in a personal way are love stories or tragedies. And those genre are rarely used in audio-visual. A few years ago a producer did make a 9-screen show on *Life and Death* which was perhaps a bit heavy-handed for the trade show at which it was seen. However, it was not a complete failure, which it would have been had the same ideas been used in a film.

To explain this particular quality of multivision, one must first consider the conditions under which it is presented. What is of prime importance is the state of mind of the average member of the audience when the show begins.

Trade exhibitions

At, say, a trade exhibition he is disorientated. There will be noise, unfamiliar people around, new products he has just discovered, people he has run in to whom he has not seen for some time. The multivision show in your stand must take all of this into consideration. The theatre in which it is seen must isolate the viewer from the rest of the exhibition. But the show must appeal to him in his state of mind at that moment. It should be fast moving, entertaining and full of ideas; it should appeal to his individual eye. It is a mistake to include too many hard facts about your products. The poor viewer is already reaching overload point through having walked around the exhibition.

Sales conferences

A quite different occasion is the sales conference. Here the viewer (it could even be the same person) is in a totally different state of mind. He is not fully an individual any longer. He is with lots of colleagues and he feels part of an organization. For the past year he has been working for that organization and now he is emotionally keyed up and waiting for the company to give him something in return. Do not disappoint him. Do not be afraid of making the multivision too idealistic in content. If it merely has a quick look at next year's work, then the audience might just as well be back in the office and getting on with it. For this kind of celebration, multivision is a superb medium.

Visitor's centre

A visitor to your company will, again, be in a totally different state of mind. Like the exhibition visitor he is not an employee and both you and your business may be unfamiliar to him. Yet he differs from the exhibition visitor because he is more likely to become a client. He has made a special effort to pay a visit and in return you will be giving him individual attention. Therefore he will be more relaxed, but also more critical, and even suspicious, unless you can win him over. Any multivision programme that you show him must be informative, arguments in it must be carefully reasoned and wholly rational and it should give a highly favourable impression of those parts of your business that you wish to display.

A multivision show on your premises should be about *your* business, not your visitor's business, nor what you presume to be able to do for him. The client will soon find an opportunity of doing business with you if he likes what he sees. A visitor's centre in a factory or office with a permanent multivision show is a window into your company. The visitor will imagine that behind that screen, in different parts of the building, tremendous activity is going on, making products or providing a service of which he had previously been unaware. A film or a video programme could make these activities appear rather ordinary but the illusion of multivision is complete. A combination of high quality photographic images and an emotive sound-track will gradually win over the most critical client.

Equipment

Consider the playback equipment which is required and the choice of media in these terms. Because film is the oldest established of the media under discussion, there are more facilities available for replay in this medium than for the others. Of 1·6 million 16mm projectors made in 45 years, 1·1 million were still in operation in 1977. So it is easy to distribute a 16mm film to a particular section of the public. But even a 16mm film does not distribute itself and far too many are left in film libraries in the hope that someone will request a showing.

Video requires a cassette player and a monitor. The availability of these begins to rival that of 16mm projectors, in numbers of units. The potential audience though, for each showing on a usual monitor is much smaller. Video projectors are less widely available.

A large multiscreen is normally a one-off installation. The show is specially made for a particular combination of projectors and control equipment. This system is erected at the viewing venue, whether it is an exhibition hall, a conference room, an hotel or in your visitors' centre. It may travel around, spending a few weeks in major cities, and it may even be duplicated. Smaller shows may be networked to a number of replay units stationed around the world. It is unlikely that you can completely avoid expense regarding the purchase or hire of hardware in which to show the programme.

Media are not interchangeable

Choose your medium with care, for its primary purpose is often unsuccessful in transferring from one medium to another.

16mm films can be easily transferred to video cassette and many of them will be acceptable; but there is always a conflict between the photographic and electronic media. What was conceived for a larger screen is reduced in replay to around 65cm (26in) and much of the detail is lost. There is also reduction in the tonal scale. But the discrepancy goes deeper than this. Film and video belong to different eras. Film-making is a craft more than a science and the presentation of a film demands a sympathetic rendering. The high technology of video reproduction overwhelms the concepts of film production. Two men with an Arriflex and a Nagra and plenty of time can make a brilliant film on nearly any subject you care to mention. It simply does not require hundreds of millions of dollars worth of investment and research in order to show the film that they have produced. Video reproduction does not reproduce the 'feel' of a film. It distances the audience from the subject. Between the viewer and the man with the Arriflex there is now all that investment and all those electronics. You could say that a video transfer of a film is equivalent to Moses making a photostat of the Ten Commandments.

This observation does concern the sponsor. It is your money and your intentions to communicate that will be disappointed.

Far worse than film production/video playback is the opposite: video production/film playback. It is technically possible but aesthetically dreadful. The reasoning behind it lies in those 1·1 million 16mm projectors; but it would be far better to make a film.

But far, far worse is the ultimate media 'crime': recording a multivision show on film and replaying it on video cassette. This usually derives from lack of foresight and commonsense combined with an uncontrollable urge towards expediency. The multivision show at the exhibition is a great success; and then the decision is taken to film the multivision and run off a few video cassettes at the same time to use, out of the country for exports. The results are disastrous. A mixing-media exercise can seriously damage the credibility of a company. It is normally done for the sake of economy but in fact the expense is almost totally wasted.

The rules of the game are these: if you make a film, project it on a film projector. If you own a video cassette player, make a video programme for it. If you produce a multivision show, purchase or hire the system for which it was designed. And before you do any of those things, make sure that you have the right medium for the job.

Costs

Having assessed the merits of each medium both in terms of what they can do and what equipment is needed for replay, the sponsor then asks how much is it going to cost? What are the relative costs of each medium? This is really a hypothetical question because one should not be comparing equally useful options for one given situation. However, one can think of hypothetical conditions where a choice may be difficult to make. Supposing a truck manufacturer wants to show a new range of trucks to the market at an international motor show. Such a major undertaking may well involve ventures into all three media. Take the centrepiece of the stand, a viewing theatre where a film, a video programme or a multivision may be shown. Each of these, in its own way, may be suitable. Each will be different and will fulfil different marketing needs. For instance, it would be possible to show on film or video how the trucks handle on the road, thus reducing the need for lengthy demonstrations and test drives. Or it may be that the company is particularly proud of their design features which will show up to best advantage on multivision. Either way, the theatre is going to cost a fixed amount and it is expected that 30 or 40 people at a time will watch the show.

Whatever happens, the project is going to be very expensive. Consider the replay equipment first.

The alternatives are:

1. A good quality 35 mm film projector.
2. Video tape recorder and projector.
3. Twelve-screen multivision equipment.

A small domestic video projector would scarcely form the centre-piece of a large stand. Roughly, the cost of installation and operation by engineers will be the same in each category. The cost of hiring the equipment, however, will be markedly different. Multivision will cost twice as much as film projection. Video will be nearly double the other two put together.

However, the replay equipment cost is only a fraction of the total budget. Say, since the equipment will all be on hire for two weeks, the following proportions would apply, (excluding the cost of engineers):

1. Film projection would cost $\frac{1}{30}$th of the production budget.
2. Video projection would cost $\frac{1}{5}$th of the production budget.
3. Multivision would cost $\frac{1}{15}$th of the production budget.

This implies that the same budget is being allocated regardless of which medium is chosen; the above proportions give an indication of equipment expenditure in terms of *relative* costs.

The amount spent on a film production would be rather more than that spent on video. It would bring the budgets for equipment and production to around the same totals. However, for a twelve-screen multivision, a really spectacular show, the overall budget could well be less, not much less perhaps, but a saving of say 20% in £50,000 or $100,000 is a substantial economy.

Of course, you cannot consider the expense until the clients' wishes are known and you have estimated the cost of a particular programme. Sponsors have to find out how much it is all going to cost and the best way of doing this is to view a number of recently made programmes and ask how much the budget was in each case. You will soon get the feel of what a $10,000 multivision looks like, and a $20,000 version, and so on up the scale. When you watch the programmes, however, take note of any factors which might distort the figures. If some of the shots have been used in magazine advertisement or press hand-outs, then perhaps only 50% of the actual cost of the photography is budgeted for the multivision presentation. Therefore the price would have been kept artificially low. Library shots and stock sequences are the producers' usual way of keeping the costs down.

Choosing a producer

Having estimated how much it is all going to cost and having settled on multivision for his particular application, the sponsor then looks round for a producer. The relationship between the sponsor and the person he commissions to make the show is important. In practice, both of these entities may be more than one person. The 'sponsor' will have many colleagues who have perhaps an equal amount of power and influence over what publicity the company should produce. There is a grave danger here of a 'committee voice' emerging. And you can not make multivision shows by committee.

One person in a sponsoring company should be appointed to the task of drawing up a short-list of potential producers. This person should have a casting vote in choosing the producer should opinion be divided after they have given presentations. He should also be the person who does the day-to-day liaison with the producer and who is also responsible for helping to explain the producer's ideas to other members of the sponsoring company. For, make no mistake, it is extremely hard to visualize the finished product from a script or storyboard. It is very doubtful whether as many as half a dozen members of a sponsoring company would have either the time or the patience to really understand a good script. But it is very helpful if one of the members can say, 'Listen, I spent yesterday evening reading this and it's just what we need.'.

Perhaps it is for this reason that many of the most successful producers are those who can explain to their clients exactly what they intend to do. Unless there is understanding on both sides at the very beginning, the gap will progressively widen until it is too costly to bridge.

In making a choice of producer the sponsor should be making a completely rational decision. It is true that many of these decisions are in practice swayed by personal friendship or acquaintance which in itself is no bad thing. If someone you know is involved in the production then there is already a basis of trust and understanding. However, merely because a friend introduces you to someone who has something to do with multivision, that is no reason in itself for engaging him to commission a script. Before you know what has happened, production is under way and you begin to realize that there are lots of other people who would make a better job of it. But by then it is too late and the casual acquaintance has established himself as yet another media consultant/producer, with your own name on his short but impressive list of clients.

A visit to the producer's offices and studios is essential if you are to judge how he works. There will be lots of tell-tale signs which will let you know whether the company is the sort to which you could safely entrust the making of your programme. The demonstration theatre should be the most comfortable and streamlined part of the office. The photographic areas should be spotlessly clean and there should be some sign of activity indicating that he occasionally does some work. Do not be put off by pieces of equipment lying around the workshop or by the mass of cables and projectors heaped up in the corner. A multivision workshop looks chaotic; it should do. This will indicate that your producer is not afraid of the tools of the trade and that someone there actually knows what the insides of the equipment looks like. One or two production companies are reluctant to use a screwdriver or a soldering iron; this is not particularly helpful if your system needs attention.

If you visit several production houses you will soon discover how varied they are. There are companies specializing in producing single screen marketing shows. Others who like to tackle exhibition programmes and product launches. There are companies who exclusively concentrate on conference production. Still others who specialize in making audiovisual shows for museums and art galleries. You may notice, too, that many producers have associate or affiliated companies who offer a further range of services. There could, for instance, be a firm of photographers in the same studio or a colour processing laboratory. The company whom you have approached could be specialists in slide photography and duplication. Or it could be a design company, or principally a film and video company. You will have to observe carefully, ask a lot of questions, and judge them fairly.

If you have a very limited budget, you might choose a freelance producer who has no overheads but does have a good reputation. Ultimately it is the producer's reputation which is your chief safeguard. A multivision producer may well live out of town where life is more congenial and still find it possible to be working full time on a number of different projects. Many of the processes involved in creating a good programme, such as research and script-writing, photography and graphics, require neither masses of equipment nor expensive premises. For those parts of production which do require expense, the producer can hire or rent specialized services such as sound recording or registration photography. However, if he makes any money at it, it will not be long before he rents a studio, if only as a tax loss.

Contracts

When you have found your producer, there is the question of a contract. This is to protect both of you. A contract guarantees that you have a show on a particular date, whatever happens. It will probably also state that your approval is necessary before a particular version of the commentary is recorded. In return you will agree that the producer has full creative control and access to any information that he requires. Principally, however, the contract will define the stages at which approval is to be given and the times when payments are to be made.

With a multivision show there are no 'rough cut' or 'fine cut' stages which you get with film production. The approval stages are: 1, script outline; 2, script or storyboard; and 3, the finished show. With a large multivision the final approval demonstration will probably not show the programme in all its glory on a large screen in ideal viewing conditions. It will be seen in a temporary theatre using a smaller screen. However, it should be sufficiently impressive for you to judge its content and technical excellence.

In between stages 2 and 3 there is often too long a gap. The member of the sponsoring company who has been delegated to look after the production should keep an eye on its progress throughout; but all those who have a rightful say should approve the programme at the three stages.

Stage 4 represents the final setting up of the show, on site, to your satisfaction. Payment is normally expected at or around completion of each of these stages. If so, the total budget is divided up and paid over a quarter at a time. More often it is divided into thirds.

The format of the show

A producer has been chosen so now return to the actual concept of your show. You should be clear about the choice of multivision formats that exists. The many shapes and sizes that multivision can assume are described in some detail elsewhere in the book. Do you need a 20 minute 3-screen show or a 10 minute 6-screen show? Do you want to see fast animated effects using three or more projectors on each screen? Or do you want to include some cine projection and programmed lighting? In fact which format is most suited to your application?

It is essential, here, to imagine the eventual viewing conditions. If you intend to have a large audience it is naturally appropriate to increase the number of screens. Of course, you can merely increase the *size* of the screens. This means that the visual content of the show is proportionately decreased. Individual screen sizes should be kept to a dimension which is suited to slide projection. This is a matter of taste and my own feeling is that 3×3 m $(10 \times 10$ ft) is a maximum individual screen size even for the largest audience. Beyond that, you should increase the number of screens. Keeping the screen size down means that conventional slide projectors with 250 watt lamps can be used.

The most attractive multivisions are those which have a number of screens, say 6, 9 or 12, which can all easily be seen by each member of the audience without undue discomfort. It is annoying to miss what is happening on the left-hand side of the screen because you are watching the right-hand side. Each person should be able to see the show as a whole. Smaller individual screens mean a higher light intensity and a relatively superior quality image.

Most subjects can be treated in different ways and there are no laws about which format to use. A show for a museum on art history would not use too many distracting animated effects, they would serve no purpose. A museum system therefore is unlikely to use more than two projectors per screen. Travel programmes designed for the tourist are particularly suited to the wide-screen format that 3-, 5- or 7-screen multivision can provide. The choice between 3, 5 and 7 should be influenced by the shape of the auditorium and the size of the budget. Beyond such simple, commonsense guidelines one can only advise that you choose a format which will meet your own

187

expectations. I read recently that a producer had convinced one of his clients that 6 projectors were just *not* necessary. 'We managed to talk him down to two.' That producer is talking himself out of business—a 6-projector show is so simple technically that it can all be operated by a 'black box' unit with a tape deck, amplifier, decoder and automatic reset built in to it. There is no need to be afraid of the complexity of the medium. Considerations such as portability and cost may enter into the argument on the 'keep it small' side.

Production time

When agreeing the format with your producer, consider how long the production will take. One of the reasons for reading this book is to find out that in advance. You may, for instance, have contacted your producer too late. Whereas you were hoping for a large 9-screen multivision for that European exhibition you now discover that there is only time to make a single-screen show. A responsible producer will not undertake work which has to be rushed through in order to complete on time. However, it is quite frequent in the multivision business to be programming early in the morning of the opening of the show. This is normally the result of last-minute changes requested by the sponsor.

Ideally all multivision shows need as much time as possible. Time is the most precious of all commodities and in professional work it is not given freely. The great advantage that the amateur or the student has over the professional is far more time to spend over getting every process exactly right regardless of how long it takes. But in professional production each stage has to be completed on schedule otherwise the whole project is no longer valid. A one-day conference or a short exhibition could be over while the producer is still seeking perfection.

So compromises have to be made and the professional producer must be judged on his ability to produce a high quality product within a given time limit. This ability depends to a large extent on the number of people whose services he can call upon. It also depends upon a degree of cooperation from the sponsor. If you delay production while colleagues are making up their minds on a particular point in the script—or while someone delays obtaining essential information or is slow over giving access for photography—then you can hardly blame the producer for running behind schedule.

From the initial conception of the show a production company should be able to come up with an outline within a few, say 10, days. A script will take about 3 weeks. These are minimum figures for the average length show. The outline, or 'treatment' will contain some basic ideas about the sequences and effects which will be used and it will say how the subject matter is to be treated. The script will have any commentary written out in full and it will have a description of the photography and effects. However, the script is just a basis to work on. The creative process has only just begun and time must be allowed for thought and reflection, brainstorming and invention, even after the script has been completed. If this were not the case then it would merely be a mechanical process from that stage onwards—and quite easy for me to quantify the amount of time required.

On an audiovisual production a photographer will expect to shoot between 15 and 30 slides a day. To achieve this high output he may in fact be using 15 or 20 reels of 36-exposure film. No doubt this will horrify some photographers who are used to working on magazines. But a film-maker would expect to get around 2 minutes of film 'in the can' by the end of the day. Multivision photographers have to be encouraged into the same attitude. Their work is now measured in different terms.

Leaving aside any process photography that has to be done in order to achieve special effects, the length of a shoot will be around 3 weeks for a short (12 minute) 3-screen show. A good rule-of-thumb would be one week for every 10 minute/2-projector 'sequence'. With large multivisions this ratio can be decreased because many duplicates will be made to show identical images on different screens, and more composite slides will be used (one picture across all the screens).

Many of the processes of production will be done simultaneously. While the photographers are working, artists will be drawing cartoons, graphics, titles and other artwork. These generally take longer than the true action photography itself. This is why when you visit your producer to see the slides set out in sequence on his light box you will notice a few gaps. 'Oh, we're waiting for those. That's why I've put blanks in for the time being.' If there are to be a lot of graphics, do not be surprised if 80% of the slides are still blank after the first couple of weeks!

Obtaining all the special slides, shooting artwork, rephotographing parts of slides, adding colour, will all take more time. Recording sound effects, mixing tracks, recording commentary and encoding will add a further week, or so, to the production time.

To recap, the following is an estimate for a 12-minute 3-screen show, using six projectors:

Outline	10 days
Script/storyboard	3 weeks
Photography	3 weeks
(Artwork	4 weeks concurrent)
Process photography	2 weeks
Recording/Mixing/Encoding	2 weeks
Total	12½ weeks, approx.

To this should be added a further 2 weeks for client approval and viewings. Thus, the production of a 3-screen show takes between 3 and 4 months. It may be possible to do it much more quickly depending on the current workload of the company involved. But remember, they will almost certainly be producing other shows at the same time and this brief description of the production process above necessarily leaves out many of the details.

Good planning and adequate warning will always pay good dividends. What matters, above all, is the final finished product—that is what the audience sees. As a sponsor you will be involving yourself in one of the smaller offshoots of showbusiness. You will be helping to encourage and develop a new medium and you will be giving your producer the means of developing the language of audiovisual. With the good will, hard work and creativity that is lavished on a multivision programme you will find that it repays you many times over, from the very first showing.

A textbook description of audiovisual production is insufficient to give the newcomer a true insight into what actually happens when a company is commissioned to make a programme. There is always a wide gap between theory and practice especially when many busy people are involved on a single project. For this reason I include a short description of an actual production, with an indication of when each stage took place.

Single-screen production: a day-by-day account

This programme has been chosen almost at random since there are many professional production companies making single-screen shows for industrial clients, and any one of these shows should reach a high level of quality and effectiveness. And yet even a description of a particular programme will of necessity leave out mention of other activities that are happening.

Martak Ltd., a UK audiovisual company specializes in single-screen productions; most of their clients are industrial (manufacturing) or commercial companies who require a professional audiovisual production service, that is, high quality work within a given time. The deadline is all-important.

This was true of a production made by Martak (Western) Ltd., who are based in the country town of Cheddar, for C & J Clark Ltd. one of the world's leading shoe manufacturers. The purpose of the programme was to present the Chairman's Annual Report to the employees, a type of application to which audiovisual is particularly suited. The shareholders meeting was to be the first public appearance of the programme. With twenty manufacturing units in the UK, and other factories in the USA, New Zealand, Australia and South Africa, the company would get excellent mileage from the programme. It would help good internal public relations by presenting the achievements of the company in an entertaining and easy-to-understand format. This is what happened.

Day 1 The telephone rings at the producer's offices. It is a potential new client, in fact, the Public Relations Manager of C & J Clark Ltd. A meeting is set up between the PRM and the production company's Managing Director, Bob Tudgee.

Day 3 At the first meeting, which takes place at production company's offices, the PRM views some programmes, likes what he sees and requests an estimate for producing an audiovisual show of the Chairman's Annual Report. The producers agree to this and request further information about the exact purpose and content of the show.

Day 5 Second meeting, at which full brief is given, enabling producers to go ahead with a treatment and an estimate.

Day 12. Full treatment and an estimate are forwarded to PRM for approval.

Day 15. Approval for both treatment and estimate is given. Client requests that the producers proceed with a draft script.

Day 22. Meeting between producers and client to discuss the draft script. Some general alterations are agreed and the OK is given to proceed with the final script. Arrangements are also made for the photography. Visits are arranged to take pictures within the local factories. Production schedule is worked out as being a total of 8 weeks. The deadline is the Shareholders' Meeting.

Day 25. A copy of the script is sent to the Chairman, in Australia. This is necessary because he is visiting many different countries on a 3-week tour of overseas factories before returning to the UK. In fact he will be in England for only 3 working days prior to 2 weeks holiday. In this instance the production schedule must be made to fit the client's schedule.

Day 30. Final recording script is approved and production can begin.

Days 30-60 Photography on location and in the studio. During this period, day 45, Bob Tudgee flies to the USA to obtain shots of a recently acquired subsidiary, the Hanover Shoe Company. On arrival in New York he is confronted with one of the fiercest snow storms for years—14 inches of snow have made transport extremely hazardous, let alone photography outdoors! With only one week available Bob flies south to get shots of the factories and shops, and in fact the results are found to be extremely good.

Day 55. Back in the studio, recording of the Chairman's message takes place. On the same day shots are taken of him for inclusion in the programme.

Days 56-70 The production team put the soundtrack together, searching out music, mixing and editing. Others are mounting slides and the show is beginning to shape up. Encoding is done and the show is ready for the first preview.

Day 77. Final presentation of the show is given to the Chairman. This takes place at the client's headquarters in Street, Somerset.

Day 80. Shareholders' meeting. The programme begins its useful life for the client and is eventually seen by all employees.

It was a successful programme, but as I said above many others could have been chosen. From this day-by-day description two important points emerge which may help other producers and clients. The first is that the client company is represented by one person who is familiar with the medium and who has the power of decision. This is the ideal situation and makes the producer's job that much easier. The second point to notice is that great importance is given to the script and planning stages with frequent consultations between producers and client. In this way mistakes are prevented and the blueprint is created for an effective production. It is also worth noting that unforeseen circumstances, in this case the weather, can affect a production but must not be allowed to ruin it.

12 In-house production and presentation

Throughout the world industrial companies are finding that their audiovisual needs far outstrip available budgets which can be allocated to outside professional productions. So they consider making their own multivision material.

It is unlikely, though, that anyone in the company has the slightest experience when it comes to producing an audiovisual show. However, a large company does have staff photographers, publicity writers and graphic artists. It may even have a few electronics experts. It will certainly employ people who have an ability to organize.

A junior executive is then appointed to look into the matter and to produce a report; he becomes enthusiastic about the project and he contacts some audiovisual equipment manufacturers. He will soon find that most manufacturers generally act as though there is no competition and that other companies just do not exist. The executive will therefore write a slightly unbalanced report, but nevertheless, one which gives a reasonable idea of the costs involved in producing in-house audiovisual shows.

There can be problems

The element in his report which is most likely to be inaccurate is the estimate for time: personnel time. This will not worry anyone unduly since the publicity department has to be paid anyway. What will not have been recognized is that the first big production is going to cost in real terms considerably more than buying the most professional and creative services that could possibly be obtained. Simple processes that take a professional a few hours will take in-house personnel several days to complete. An expensive secretary will spend all day mounting slides, without the help of special dust-free loading bays or dust-extracting brushes. The slides will go out of sequence and have to be painstakingly numbered. Then it is discovered that the stick-on numbers are helping to jam the projectors and they all have to be removed —after first finding out the cause by asking a helpful manufacturer. Then someone notices that the slides are too dirty and the whole process starts all over again.

Another element in the report which will be inaccurate is the absence of any reference to quality. The executive will not suggest the truth: which is that it is impossible for the in-house production to come anywhere near to a professional production because the amount of equipment needed to produce such a show is extremely costly. He cannot, for instance, suggest that the company purchase an eight-track sound studio. After all, the staff photographers are probably only issued with a fraction of the equipment owned by the average professional with his own business. Neither will the executive suggest that the company should employ a scriptwriter and a sound recordist —two of the most important people in an audiovisual production team.

When the report is submitted to the board it will state, in so many words, that instead of spending, say, $12000 on a professionally made show and, say, $3000 on equipment to show it with, there is a recommended alternative. Namely, to purchase the means of making an unlimited number of productions for approximately the same outlay. In far too many cases, companies go ahead on that basis only to be disillusioned as time and again equipment does not function correctly owing to inappropriate operation and incorrect production. Programmes are made, which, while having all the knowledge of the company's experts poured into them, just do not communicate it to other people. Each programme becomes a succession of still pictures— highly suitable for a brochure—accompanied by a badly recorded commentary. The end result is that the whole enterprise is shelved, literally—with very valuable audiovisual systems gathering dust in company storerooms. The unlucky executive who was appointed and became so fired with enthusiasm is now in danger of being simply fired.

The right approach

So, what steps should be taken to ensure that the whole enterprise does not end in failure? It should certainly be made clear that any undertaking that involves an important public exhibition of the company's image should be preferably commissioned or, at the very least involve a professional scriptwriter and producer.

Product launches, conferences, exhibitions and marketing programmes fall into this category. However, in areas such as staff training, internal communication, internal information on job opportunities, pension schemes, company growth, labour relations, there is no reason why skilful and effective programmes cannot be tackled in-house. The same presentation equipment can be used for both commissioned productions and for in-house productions. Emphasis should certainly be placed upon obtaining the best presentation equipment that the budget will allow.

Realistic planning

The executive's report should make the point that the audiovisual industry, which is growing with great speed, is extremely diverse in its structure. Not only are there companies who will undertake a whole production from briefing through to a finished programme, but there are equipment and studio hire companies, sound studios with expert staff who will assist a novice producer, music libraries with thousands of pre-recorded backing tracks and even companies who specialize in encoding audiovisual shows.

Clearly all the facilities are available in most countries but each time a new customer appears he has to research the field. There are many publications, mostly magazines, but few books which will give technical help and point the newcomer in the right direction. Unfortunately, some of the professional help of the kind to be sought out, is not always as experienced and creative as one might hope. Anyone purporting to be a producer or scriptwriter should have his credentials carefully examined before he is given even the most humble production.

It is feasible, therefore, to cost out how many of these facilities will be required during the first year and how much actual production equipment will be needed. The report should also state quite clearly and honestly how much time will have to be spent on audiovisual production by company personnel if the target number of programmes is to be achieved.

Basic requirements

First of all outline the minimum requirements and most effective methods of producing a short-single screen, two-projector show in-house. If this is the first production of several on a number of different subjects the purchase of equipment is very important. The organizer may have to liaise with overseas subsidiaries regarding the correct playback systems which they need purchase.

Fortunately slide/tape equipment is easier to use on a worldwide basis than video equipment, with its several different standards. Even so, it is important to be sure that the correct equipment is available in the countries for which the programmes are intended.

The equipment needed for audiovisual presentations is made by specialist manufacturers. Some items have long delivery dates and so must be ordered well in advance of the required day.

Assume that the intended audiences for this show are going to number around 20–30 people at a time. However, a large presentation to, say, 200 people is possible. The playback equipment needs to be portable because it is intended to take it around to different offices and there is no guarantee about the size and shape of rooms. It should also be easy, if not foolproof, to operate.

The most convenient systems are self-contained 'package systems' that require a minimum of setting up. The single unit consists of a cassette deck, an amplifier, a decoder to unscramble the pulses on the fourth track of the tape and pass commands to the integral dissolve unit. The presentation unit controls two automatic slide projectors, and uses a separate loudspeaker. Add to them lenses, connecting cables and carry cases to complete the outfit. Additional lenses should be added to cater for projection in a very large room. 180 mm is a useful focal length. A projector stand is invaluable for immediately providing a firm base in which to position the unit. It should be easily portable and should stand rock steady with a top shelf height of around 150 cm (5 ft).

The final item is a portable screen. There are many screens on the market, many very good. A tripod screen, though, is strictly for the amateur. It never seems to point in the right direction. The sides of the screen curl up after very little use and the whole object is more reminiscent of the classroom than of an efficiently run business. To be preferred are the types of screen that consist of a lightweight aluminium frame that zig-zags out to form an exact square or rectangle with corner supports. On to this you clip either a front or rear projection screen that has to be stretched into position. Then two T-shaped legs are bolted to either end, and the height can be adjusted during this last operation.

Here, then, is a list of equipment for playback of the programme, given the conditions set out above:

1. Presentation unit with cassette deck, decoder, amplifier and dissolve unit built-in.
2. Carrying cases
3. Loudspeaker
4. Variable focal length lenses (zoom lenses)
5. Long throw lenses (180 mm)
6. Loudspeaker connecting cable
7. Remote control and extension cable
8. Two automatic slide projectors
9. Projector stand
10. Screen
11. Spare lamps and fuses

If every department to which the material will go has this equipment, then our producer's work will not be in vain. The programme will be played, maybe hundreds of times to thousands of different people.

Production equipment

The next step is to look ahead to see what kind of production equipment will be needed. Assume that the staff photographers have all the necessary cameras, lights, etc. that are needed to take good colour transparencies. However, since the show is being sent to a number of locations, you need a number of duplicates. Should these be purchased outside or made within the company?

If the programme is to have only about six copies made of it, why not shoot more originals? This is assuming that the original photography is being taken on 35 mm slide film. Of course, if the photographers are going to shoot dozens of experimental reels in the hope of getting some good shots it would be foolish to tell them to take each picture six times. However, when they can control everything it is quite feasible to make six originals instead of one. After a few test rolls, they should be able to keep bracketing to a minimum.

Some professional producers disagree with this suggestion but they are probably far better equipped for making duplicates than most in-house studios. Not only that, but a duplicate can never match the quality of an original. In the last analysis, it must all be costed out. The professional producer never knows when his client is going to ring him to order another couple of copies of the show and this method is not therefore recommended to outside studios. If however a hundred copies are needed then duplicates must be made and appropriate machinery bought for the purpose.

Registration photography

Another necessity is some means of reproducing artwork, ideally a registration rostrum camera. The professional producer can spend the price of a house on this single item. Obviously the in-house unit cannot justify this type of expenditure. For those sequences that require registration slides— for instance showing the Chairman's face break into a rare smile without it being obvious that two (or more) different slides are being used—you have to go to outside services. Simple document copying, though, should be well within the capabilities of an in-house team. The best compromise is to set aside an area where flat artwork and small objects can be photographed and where correctly balanced lighting is permanently in position. Luxuries such as polarized light or electrostatic systems for holding artwork flat will have to be foregone. If possible an accurately registering camera with a high quality close-up lens should be mounted vertically above an adjustable frame. It may, for instance, be possible to purchase a secondhand rostrum from a film animation studio—or perhaps to adapt a photographic enlarger.

The most valuable asset is office and work space. When the equipment starts to arrive there must be an area where it can be safely kept, preferably without having to pack it all away into a cupboard.

Sound recording

Commentaries are normally recorded in a cramped soundproof kiosk with a chair, table, microphone and a glass of water. It should, though, be made slightly more comfortable; because not only the managing director but possibly the occasional TV personality may be called upon to spend an hour or so inside it.

However, now is the time to consider employing a professional sound studio for the final recording.

Before tackling any in-house recording, you should practice with sound equipment before risking complete failure with recording levels, cross-talk and other technical hurdles, in the presence of the boss. The basic equipment is a 4-track reel-to-reel tape recorder and a really good quality microphone and stand.

Audiovisual presentation systems often use what is known as 'audio-visual format'. That is, they take an ordinary cassette but use the full width of the tape in a single pass across the playback head. The cassette player in a presentation unit does not record. This is an advantage because eventually someone in one of those overseas subsidiaries would otherwise accidentally wipe both the audio and pulse track off the cassette. So, what is needed is a cassette recorder that reproduces a cassette to the same format as the one being used for playback. The cost of this recorder is about one-third of the cost of the reel-to-reel.

The next item on the shopping list is an encoder: a small box with a row of buttons for generating pulses that will trigger off dissolves between projectors. Some dissolve units and presentation units generate encoding signals, and thus obviate the need for a separate encoder.

An encoder, though, has an extra use. It can decode as well. This is ideal for presentations in larger rooms (seating, say, 60 or more) where the presentation units of 10 watts are rather too small. Ideally it is nice to achieve a really excellent quality in the audio playback at headquarters. Especially since it is to the main conference room that many important visitors will be shown and it is one of those venues where the company's image is definitely at stake.

Really good quality audio systems tend to be bulky, and portability can be an important criterion in selecting systems for overseas subsidiaries.

Having purchased an extremely good 4-track reel-to-reel recorder for production, you might just as well make use of it for presentation. For this, you need a twin-channel amplifier which will amplify stereo signals, rated at around 60 watts per channel. To complete the system, you need a separate dissolve unit which is *identical* to the one that is integral to the presentation

198

units. It is most important that it is identical because it must reproduce the same visual effects that can be obtained.

The first showing

Early in the project, it is important to imagine the first screening of the show. Is the conference room really a suitable venue? Many boardrooms and conference rooms have cinema screens built into them. Others are oak-panelled and expensively furnished so that any introduction of audiovisual equipment appears to be an unwelcome intrusion. Assume that there is a front projection screen, with elegant curtains, and a small projection room at the other end. Of course, the window to the projection room needs enlarging so that two-slide projectors can 'see out' properly. Motorized curtains and a dimmer for the houselights can help to create the right conditions.

The best way to mount the projectors is on registraction stacking stand to hold the projectors in a rigid, yet fully adjustable position, one above the other. This reduces the keystone distortion on the screen to an absolute minimum.

Here is a list of the audiovisual equipment needed for a high-quality production.

1. $\frac{1}{4}$ inch 4-track reel-to-reel tape recorder. (Small monitor amplifier and monitor speaker, optional.)
2. Encoder/Decoder for single screen shows.
3. Dissolve unit.
4. Amplifier rated at about 60 watts rms per channel.
5. Two loudspeakers. Sited near screen.
6. Two automatic slide projectors.
7. Registration stacking stand.
8. Cassette recorder (with audiovisual format head)
9. Microphone and stand.

A good start

Started this way the audiovisual project has a very good chance of succeeding, of course there are limitations and outside services are important. If the first few programmes are successful then there is a good chance that the company can justify the expense of permanent technical staff and the purchase of extra equipment. Nothing suggested above will be redundant if this happens. In the future, when complex soundtracks can be recorded, other sources of sound will be required: another tape recorder, another microphone, a sound mixer, and a record deck.

Starting out

Of course, there is no point in buying the equipment in isolation. Normally, the best person to organize that is the man responsible for producing the first presentation, and to do the two things at the same time.

As for a magazine article or a documentary film, the subject must be thoroughly researched. The subject in this case is a new chemical plant; and the reason for making the programme is to inform overseas subsidiaries about the new process in such a way that the audience will really understand it and even feel what it is like to operate the new machinery. Only a good audiovisual show can do this kind of task without excessive expense.

The man in charge of the show is, say, assistant editor of the company journal—call him the producer. His first job is to find out about the new plant from its designers. After his meeting with them he should not have agreed that one of the designers will 'come up with a script' in a few weeks' time. That spells certain death right from the start. Rather, the producer should come away with stacks of written information and a clear idea of the unique points about the new process. He should also have a good conception of what the designers would like to see in the show. This he should find out— but, of course, it is not their programme and in the interests of communication their ideas can, in the last resort, be gently ignored.

The next step will be to visit the new plant and take careful note of each activity in it, talking to the workforce as necessary. A journalist will be used to doing this in any case, but the relevant areas need careful appraisal with a film director's eye—not a photographer's eye. A couple of striking shots of a process may be sufficient for the company newspaper but they could completely disorientate an audience.

The producer should be able to tell how much photographic equipment is needed to get a really comprehensive series of shots. He should be able to gauge the amount of artificial lighting needed and tell whether all this equipment is going to disrupt the normal working of the factory. He must also try to pick up tips about what not to shoot. (For instance, the portable washroom facilities indicating that the plant is nowhere near completed.)

During the visit the producer should be looking for ideas—not too hard at this stage because it is very easy to start off with a facile idea and hard to reject it once you have woven a script round it; but he should be looking. For instance, there might be an attractive and intelligent girl who could take part in the show and add a touch of human interest. Or the foreman may turn out to have an interesting speaking voice. Incidents like this could be lost if the producer is not trying all the time to find positive contributions to the forthcoming production. It is often a good idea to take instant photographs at this stage. This makes it easier to explain to a photographer about the kind of pictures he needs. They also act as a useful visual memo when conceiving the script.

200

Scripting

With a check that the hardware for the project is correct, the producer can proceed with the software. He has reached the end of the research stage and has begun to write the script. This is the most important creative stage of the whole project. The audience must be entertained as well as be informed. The sequences that you create must have a structure. A good script is like a good joke—it has the same elements, even if you are not trying to be funny. The technique is to set up certain anticipations in the viewer—but do not disappoint him or even fulfil all the expectations. Surprise the viewer, that is the secret, and keep on doing it.

The best length for the first show is about 120 slides (60 in each circular tray) lasting about 12 minutes. Longer shows with magazine changes after about 20 minutes are likely to bore the audience. The art of producing an audiovisual show is to say things as quickly as possible. Learn the technique from television advertisers—see how much they can get into a 2-minute slot. Note too, how concise and accurate TV commercials are. The message is very rarely confusing. You are not left with the impression (for instance) that tigers drink petrol or that only chimpanzees should drink tea. Who needs 30 minutes to outline the major points about a new chemical plant?

The trouble with long programmes is that they debase the value of each sequence; 10 minutes of audiovisual is really the equivalent of a 1-hour lecture with no visual aids. It is intensive learning and it should not be made diffuse.

The producer, with his scriptwriter's hat on, should therefore observe the first rule: keep it short. The next rule is less obvious: vary the pace. A programme needs an introduction to allow people to settle down and adjust to the viewing situation. So a gentle beginning is a good idea. Alternatively, you may want to make a big impact at the beginning to wake everybody up. The following would make a good introduction for the programme on the new chemical plant.

VISION	SOUND
1. Black and white picture of old industrial building	"Dirt . . .
Snap change to	FX: FACTORY NOISE
2. CU of old generator	Noise . . .
Snap change to	HALF FADE FX.
3. BCU of smoke from plant	Fumes.
Fast dissolve to	
4. Another angle of old building	Scarcely the ideal working environment
Fast dissolve to	
5. Exterior colour shot of headquarter's office, showing Chemicals Ltd. sign	And one which the board of Chemicals Ltd. were determined to change.

Fast dissolve to
6. Exterior colour shot of new plant   FADE UP MUSIC.

The script does not have to be typed in this format but it must carry a certain amount of information. A short description of the slide should be on the left-hand side of the page. Each slide change should be numbered. The type of slide change should be indicated. In the column headed *Sound* the commentary should be in lower case type. All sound effects (FX) should be in capital letters to distinguish them from the commentary. Other instructions should also be in capital letters.

This is the version of the script that will be read by a commentator; so it must be neatly presented. It is also the version that will be submitted for approval. The many different stages that a script can pass through are covered more fully in the chapter on scriptwriting. There may be a treatment, for instance, where the writer simply notes down his intentions in terms of sequences and aspects to be covered. And after the scripting stage, a storyboard may be drawn up in order to explain the concepts more easily to both the photographers and the in-house staff who have to give the go-ahead.

There are no indications in the above example of what type of music is to be used or exactly what sort of sound effects. Apart from CU (Close-Up), BGU (Big Close-Up), 'black and white' and 'another angle' there are no exact directions to the photographer. In other words, it assumes that someone is actually going to direct the programme. And if this person is the man who wrote the script, the one-man-band producer, then he should know what he wants.

There are also no timings on the above example. These will be worked out later. The commentary itself, a normal speaking speed, sets the pace. But it is very important to remember all the time the particular limitations of the playback system. For instance, in the presentation system we have described there are just two slide projectors. Each one needs time for the dark projector to advance its slide into the gate. This takes approximately 2·5 seconds. To be on the safe side, allow 3 seconds between cues when using the fastest rate of dissolve. With a slower rate of dissolve being used, this time should be lengthened accordingly. The tables in the Appendix may help you to work this out.

Photography

On this occasion the producer decides that a storyboard is not necessary. He has approval to make the programme on the strength of an excellent script and instructs the photographers to get out and get the right shots.

Audiovisual photography calls for its own techniques. Perfect photographic technique is less important than the overall impression. After all, each slide will be seen for only a few seconds. Of course, technical faults are easily noticed and I am not suggesting that photography for audiovisual needs less care and attention. But each image is not a work of art in itself. It does not need the kind of excellence that is obtained by spending all day setting up a single shot, adjusting the lighting, building light tents and so on. The whole programme is the finished work and each image must contribute to its overall quality.

Audiovisual shows need more dramatic photography than magazines or catalogues. The photographer should read the script over to himself while taking the pictures and try to see the finished work in his mind's eye. Although it will never turn out quite as envisaged, partly because other people's ideas and work are involved at least it means that the photographer is working in the right medium.

The images in a single-screen show follow each other without a break. They must therefore be arranged in a sequence that observes both content and form of succeeding slides. For instance, if you follow a picture of a chemical process with a close-up of a typical canteen lunch you must expect the audience to react badly. There is an association of ideas that may not be intended. The food will look (feel) contaminated and artificial, however beautifully appetizing the photographer may have made it appear in his individual picture.

That example refers to the 'idea content' of the slides. The 'formal content' is an abstract one referring to verticals, diagonals, horizontals, shapes and colours. If you place two pictures in succession, taken by the same focal length lens, that show the horizon line at different heights, then again the audience can feel uncomfortable.

Neither our new producer nor his photographer will know all these 'rules'. They should certainly not let that stop them from going ahead. There are so many rules—many of which can be broken—that they are learned by experience and become 'second nature'. However, if they are interested they should study audiovisual shows that have been made by other people and criticize them carefully.

Graphics

While the location photography is in progress the artists can be working on the graphics. Nearly every audiovisual show has some graphics in it. Some of them have nothing but graphics. The production of artwork is time-consuming and therefore expensive. There is so much more information in a photograph of the real world that graphics can be boring unless used sparingly to make a particular point. Facts and figures often need visual reinforcement and they are frequently easier to explain visually. However, if your audiovisual show is entirely about statistics, it may be better to send round a three-page memo instead. It is easier to read figures on a printed page, probably because it gives them a little bit more material reality. On the screen they look insubstantial and it is likely that the members of the audience will vary greatly in their ability to do mental arithmetic.

The most effective graphics are cartoons and simple diagrams. Never reproduce a page of figures in black-and-white. It will be psychologically unreadable. It is also trying to the eyes to see 99% of the picture as brilliant white reflected light while attempting to read the black type on top of it. Rather, if you must have script, reverse the image; project light-coloured figures with a black background.

Artwork will frequently need to be designed and shot in register so that graphs and charts appear to grow as in an animated film. Using cellular overlays makes this quite straightforward. Instead of drawing up a single piece of artwork, make a composite, consisting of a number of celluloid sheets on which each stage is drawn. These are then photographed on the rostrum, first one sheet, then two, and so on. The resulting transparencies from this work are then mounted in register slide mounts.

A register slide mount will hold a piece of 35 mm film in an exact position. It has at least four plastic pins inside it, in fixed positions, to prevent any lateral or vertical movement of the slide. Such mounts cost many times more than ordinary mounts, and should only be used where necessary.

Handling the slides

When the location photography is finished, the rolls of film will be examined and all the useful pictures, far more than are needed, will be mounted. Although register mounts are unnecessary throughout, it is essential that a good quality plastic or metal mount be used. It must also contain thin windows of glass to hold the slide in the same plane when it is projected. Multivision projection does not have the facility to adjust the focus of each slide. The heat from the projector lamp can alter the focus if cardboard mounts or glassless plastic mounts are used. The slides are selected in sequence on a light box—which is simply an opal perspex sheet lit from below.

Putting the show together

While the producer is waiting for all the slides to be finished he can continue with the soundtrack. He may already have recorded a few sound effects with the cassette recorder, providing it has the facility for microphone input. Remember that the primary purpose for which it was bought was to make cassette copies for the presentation units. These sound effects recordings he puts to one side for a moment and makes a trial run at reading the commentary himself. This is recorded on track 1 of the reel-to-reel tape.

When all the slides have been prepared they are loaded up into trays. The trays are numbered A and B for the left-hand and right-hand projectors (as you look towards the screen). The loading begins with position 1 in the A projector. Slide 2 goes into position 1 on the B projector—and so on. Blank slides are placed into the gate of each projector. Blank slides are also added at the end of the slides in each tray.

Now the producer draws up a cue sheet. This is a version of the script with exact timings on it, in minutes and seconds and even half seconds on occasion. The speeds of dissolve changes are clearly shown. Of course, all must be within the capability of the equipment.

The show can now be temporarily put together to see if it works. Even when you intend to obtain the services of a professional commentator and sound studio, a run-through is a good idea. You can then easily see if more original work has to be done: script changes, new photography, etc.

Encoding

Encoding is simple once you have sorted out how to plug the system together. In this example we are using a pulse-type method of encoding. The programmer (encoder/decoder) generates two different frequencies. These are quite audible if amplified. The tone bursts are of a measured duration. There is a long and a short pulse of the same frequency and a single pulse of the second frequency. This enables three different commands to be given to the playback equipment. For example, it could mean two rates of dissolve and a shutter-operated (hard cut) change.

The output of the encoder is connected to the track 4 line input of the tape recorder. In order to see what is happening the producer connects the monitoring output of track 4 back into the decoder part of the programmer. Finally a link is run between the output of the decoder to the dissolve unit.

The dissolve unit will be connected up to projectors. All mains cables will be plugged in (how often people forget to do that on the first occasion). The monitor amplifiers for replay of track 1 are adjusted.

At the very beginning of the tape a continuous burst of set-level tone should be recorded. This set-level is not used in replay. It is a guide to make sure that the recording is not under- or over-modulated. The sensitivity of the tape recorder is adjusted according to the manufacturer's recommendations. It is only necessary to record set-level if copies are to be made.

Having done this, the producer starts programming. First he places what is known as an engineering pulse on the tape. This initiates a slide change on projector A, having dissolved to the blank slide on projector B. The next pulse that is put on the tape will fade up the first slide of the programme.

If a mistake is made during programming it is easy to go back and erase the pulse. Take care that you are not erasing the commentary as well.

At the end of the session the producer will have his first audiovisual programme. Many in-house producers never get beyond this degree of professionalism. However, because the medium uses a succession of unmoving images it is necessary to compensate by having a lively and dynamic soundtrack. The human voice alone cannot supply this. Music and sound-effects are essential.

The sound laboratory

So the producer goes to a professional studio, where, within 30–60 minutes, a commentary is recorded to his satisfaction. A separate music and effects track is made, most of the morning having been spent listening to library music and selecting appropriate tracks. This separate M & E track is useful because a foreign language commentary can be made at a later date without having to remake the whole soundtrack.

Finally the commentary and the M & E track are mixed down and transferred to the producer's format which is, say, $7\frac{1}{2}$ ips on tracks 1 and 2.

Back at the office the producer uses this transfer tape for re-encoding. The practice gained in the previous session proves to be invaluable. Mono cassette copies are made for the patient overseas subsidiaries and everything is ready for the premiere.

If all the above advice is taken you can be sure that the premiere will be a great success. The novelty and satisfaction in producing a first programme are enormous. But it is only the beginning. If the programme is a success, people from all departments will be knocking at the door requesting, begging or demanding that you make an audiovisual show for them—then the real problems start.

The communications room

Every major company has a board room, the primary function of which is to provide a location for directors' meetings. Typically, the decor will simulate an aura of stability, conservation, wealth and tradition. Even if the office block itself presents a dynamic image of modernization, it is almost certain that the board room will be oak-panelled and hung with gilt-framed oil paintings.

Since this splendid room will be conveniently situated near the chairman's office it is likely that small parties of guests are often taken there for high level negotiations. But investment managers, bankers and delegations from foreign governments are usually more interested in what a company is actually doing than in whether the chairman has good taste in Rembrandts.

The company then goes through a perilous period when make-shift audiovisual shows are staged in the board room with tripod screens balanced under the chandeliers and directors craning their heads around those of their colleagues in order to get a better view.

Fortunately, this period is often short-lived because someone realizes the need for a special room fitted with the latest equipment for showing visual presentations in several different media. The communications room is born.

A communications room is *not* a cinema for general staff use, nor is it a training room for teaching the firm's truck maintenance engineers. It is a small, exclusive room which provides an ideal environment for about twenty people to see and hear audiovisual media. Its decor will be modern, suggesting that the company intends to be around in the next century rather than announcing the fact that it was founded in the last. The message to give to the designer is that Stubbs is *out* and microprocessors are *in*.

Designs for audiovisual

You should ensure that the designer involves himself in discussions with audiovisual suppliers at the earliest stage. Few, if any, companies are equipped and qualified to undertake both the interior design and the audiovisual system design. Be careful of giving the entire job to one firm who will sub-contract most of the work and charge double for the hardware. Rather, it is better to delegate a suitably intelligent and enthusiastic executive in your company to liaise with and organize the other companies involved.

What media are required in a communications room? I would suggest the following, in no particular order of priority.

1. Film, 16mm or 35mm projection.

2. Television. Video cassette and big-screen projection.

3. Multivision. A substantial three-screen or six-screen show, with at least three projectors per screen, and stereo sound.

4. Random Access Slide Projection. On to each of the multivision screen areas.

5. Single-Screen Audio Visual. A simple two projector set-up, independent of the multiscreen show.

I will say little about the first two media other than to mention a preference for video projection rather than the use of monitor sets. The video source should be $\frac{3}{4}$ inch cassette, interchangeable between American and European standards.

To design a room for ideal projection of these media is not a simple task. To begin with, many different screens are required and these must be smoothly interchangeable. Alternatively, more than one viewing direction can be used and the audience can be provided with swivel seats. However, this means that the floor has to be level, rather than stepped – which is the ideal that we are aiming for. Each person must have a completely clear view of each screen. One solution is to use both front and rear projection.

The video projection should be treated separately as a less spectacular and more intimate medium. A small video projector takes up very little space and it is quite acceptable to have it in the room with the audience, whereas all slide and film projectors should be concealed. Thus, the video is a side show and the screen should be close to the viewers. This differentiation in the physical depth in the placement of the media can be used skilfully by the person controlling a presentation. Changing from the big-screen to a video programme is equivalent to going from a long-shot to a close-up in a film. But the effect is ruined if the video is too remote to be able to see the eye movements of a person who has been recorded.

The big screen should really be just that—big. It should all but cover one entire wall. Likewise, the front projection screen that either slides or rolls down in front of the rear projection surface should be larger than you might think is necessary. Both the film and the single screen audiovisual should be Xenon-powered to give a bright picture over a large area.

The whole environment should be created specially for projection and for sound reproduction. Acoustics are extremely important, although quite easy to control in a medium-sized room. Sound amplification will need to be in the region of 60 watts per channel. Large, quality loudspeakers should be used and the sound systems for the multivision and for the film soundtrack should be kept separate.

Whilst it may be essential to have a qualified engineer on call during important shows, all of the systems with the possible exception of film should be controllable from inside the communications room. All the presenter has to do is to push the correct button on the control panel in front of him. This gives the impression that the presenter is in charge, which is far preferable to displaying your incompetence with the media by making hand signals to a projectionist. It also means that you can change your programme without having to warn someone else that you have done so.

The control panel

The panel is independently lit by a small light on a dimmer control. It has a manual override for the room lights, dimming, together with key-operated start switches for the one (or two) permanent multiscreen shows. It also has switches for controlling the opening and closing of curtains to preset positions and for sliding the front projection screen into place. Other controls will include the single screen audiovisual start, and manual slide advance and reverse. No button should operate unless the correct combination of curtains/screen is achieved. An LED read-out informs you what to do.

The random access projection can be controlled from a panel of numbered buttons for selecting the correct screen, projector, and slide number. It is convenient to build in a slide index which should always be kept up to date. This can be in an illuminated sliding drawer under the control panel.

Some companies may favour the use of a lectern for centralizing the controls. A communications room, though, should offer a soft environment where you can talk *with* your guests rather than confronting them from a lectern.

It is important to ensure that the physical conditions are extremely comfortable, that there is good air circulation in the room and a quiet air-conditioning system. There should be places on which to put papers or notes; and everyone should have an ash-tray.

All too often existing rooms are adapted and their bad features are retained even after a great deal of money has been spent. Certainly every new headquarters building should have an area set aside for this purpose. And older buildings can be adapted successfully if expert audiovisual help is called in at the outset.

Appendix 1

The SI unit system

1. *Basic SI Units*

Physical quantity	*name of unit*	*symbol*
Length	Metre	m
Mass	Kilogramme	kg
Time	Second	s
Electric current	Ampere	A
Thermodynamic temperature	kelvins	K
Luminous intensity	candela	cd

2. *Derived Units*

Force	Newton	$N = kg\ m/s^2$
Work, energy	Joule	$J = Nm$
Power	Watt	$W = J/S$
Electric charge	coulomb	$C = As$
Electrical potential	Volt	$V = W/A$
Electric capacitance	Farad	$F = As/V$
Electric resistance	Ohm	$\Omega = V/A$
Frequency	Hertz	$Hz = S^{-1}$
Magnetic flux	Weber	$Wb = Vs$
Magnetic flux density	Tesla	$T = Wb/m^2$
Inductance	Henry	$H = Vs/A$
Luminous flux	Lumen	$lm = cd\ sr$
Illumination	Lux	$lx = Gn/m^2$

Projection chart for Berthiot lenses

Slide format 35 × 23·4 mm

Projection Distance
(m)

Image width (m)	25 mm	35 mm	50 mm	80 mm–125 mm zoom		90 mm	150 mm	180 mm	200 mm	210 mm	250 mm	290 mm	360 mm
0·20	0·265	0·28	0·42	0·57	0·83	0·625	1·11	1·29	1·48	1·36	1·74	2·11	2·41
0·40	0·40	0·49	0·72	1·14	1·62	1·25	2·11	2·50	2·73	2·61	3·24	3·81	4·82
0·60	0·54	0·71	1·00	1·64	2·27	1·71	2·95	3·60	3·93	3·87	4·76	5·56	7·00
0·80	0·67	0·90	1·26	2·08	3·00	2·25	3·81	4·44	5·06	5·12	6·23	7·27	9·11
1·00	0·84	1·13	1·58	2·55	3·70	2·75	4·58	5·50	6·25	6·43	7·69	9·00	11·21
1·20	0·98	1·30	1·86	3·03	4·38	3·28	5·54	6·56	7·38	7·56	9·23	10·71	13·33
1·40	1·14	1·52	2·17	3·53	5·11	3·81	6·45	7·57	8·51	8·81	10·71	12·44	15·56
1·50	1·22	1·63	2·28	3·75	5·42	4·08	6·91	8·11	9·12	9·43	11·39	13·29	16·67
1·60	1·31	1·70	2·44	4·00	5·78	4·30	7·37	8·62	9·70	10·06	12·15	14·18	17·56
1·80	1·42	1·91	2·73	4·50	6·51	4·84	8·23	9·63	10·85	11·31	13·62	15·82	19·78
2·00	1·54	2·13	3·01	4·93	7·22	5·38	9·09	10·70	12·06	12·57	15·09	17·56	21·89
2·20	1·70	2·32	3·31	5·42	7·89	5·87	9·95	11·68	13·16	13·75	16·60	19·29	24·00
2·40	1·85	2·53	3·58	5·85	8·61	6·40	10·86	12·68	14·36	15·00	18·08	22·86	26·11
2·50	1·93	2·64	3·73	6·10	8·96	6·67	11·24	13·21	14·93	15·62	18·79	23·81	27·13
2·60	2·01	2·74	3·88	6·34	9·32	6·92	11·68	13·74	15·52	16·25	19·55	24·76	28·21
2·80	2·16	2·88	4·18	6·83	9·96	7·45	12·58	14·79	16·72	17·44	21·02	26·31	30·43
3·00	2·29	3·09	4·48	7·32	10·68	7·94	13·46	15·79	17·82	18·69	22·53	28·19	32·48
3·20	2·45	3·29	4·73	7·76	11·36	8·47	14·36	16·84	19·01	19·93	24·08	30·07	34·65
3·40	2·60	3·45	5·03	8·24	12·07	9·00	15·20	17·84	20·18	21·13	25·58	31·54	36·81
3·50	2·68	3·55	5·18	8·48	12·43	9·26	15·64	18·37	20·77	21·75	26·27	32·46	37·89
3·60	2·75	3·65	5·33	8·73	12·78	9·53	16·09	18·89	21·29	22·37	27·02	33·39	38·97
3·80	2·89	3·86	5·60	9·22	13·50	10·00	16·97	19·90	22·47	23·63	28·53	34·90	41·03
4·00	3·03	4·06	5·90	9·64	14·22	10·53	17·87	20·94	23·65	24·88	30·00	36·73	43·19
4·20	3·16	4·26	6·19	10·12	14·92	11·05	18·76	21·94	24·83	26·12	31·48	38·57	45·35
4·40	3·31	4·47	6·49	10·60	15·61	11·58	19·60	22·96	26·01	27·31	32·97	40·18	47·51
4·50	3·38	4·57	6·60	10·84	15·96	11·84	20·04	23·48	26·53	27·94	33·72	41·10	48·49

Projection chart for Berthiot lenses

Slide format 37·5 × 37·5 mm

Projection distance
(m)

Image width (m)	25 mm	35 mm	50 mm	80 mm–125 mm zoom	90 mm	150 mm	180 mm	200 mm	210 mm	250 mm	290 mm	360 mm
0·20	0·245	0·26	0·39	0·532	0·58	1·04	1·20	1·38	1·27	1·62	1·97	2·25
0·40	0·37	0·46	0·67	1·06	1·17	1·97	2·33	2·55	2·44	3·02	3·56	4·50
0·60	0·50	0·66	0·93	1·53	1·60	2·75	3·36	3·67	3·61	4·44	5·19	6·53
0·80	0·61	0·84	1·18	1·94	2·13	3·56	4·14	4·72	4·78	5·81	6·79	8·50
1·00	0·76	1·05	1·43	2·38	2·57	5·14	5·13	5·83	6·00	7·18	8·40	10·46
1·20	0·89	1·21	1·69	2·83	3·06	5·17	6·12	6·88	7·06	8·61	10·00	12·44
1·40	1·03	1·42	1·97	3·29	3·56	6·02	7·07	7·94	8·22	10·00	11·61	14·52
1·50	1·11	1·52	2·07	3·50	3·81	6·45	7·57	8·51	8·80	10·63	12·40	15·56
1·60	1·19	1·59	2·22	3·73	4·01	6·88	8·05	9·05	9·39	11·34	13·23	16·39
1·80	1·29	1·78	2·48	4·20	4·52	7·68	8·99	10·13	10·56	12·71	14·77	18·46
2·00	1·44	1·99	2·73	4·60	5·02	8·48	10·00	11·26	11·73	14·08	16·39	20·43
2·20	1·54	2·17	3·09	5·06	5·48	9·29	10·90	12·28	12·83	15·49	18·00	22·4
2·40	1·68	2·36	3·25	5·46	5·97	10·14	11·83	13·40	14·00	16·87	21·34	24·37
2·50	1·75	2·46	3·37	5·69	6·23	10·49	12·33	13·93	14·58	17·54	22·22	25·32
2·60	1·82	2·56	3·51	5·92	6·46	10·90	12·82	14·49	15·16	18·25	23·11	26·33
2·80	1·95	2·68	3·78	6·37	6·95	11·74	13·80	15·61	16·28	19·62	24·56	28·40
3·00	2·07	2·89	4·05	6·83	7·41	12·56	14·74	16·63	17·44	21·03	26·31	30·31
3·20	2·21	3·07	4·28	7·24	7·91	13·40	15·72	17·74	18·60	22·47	28·07	32·34
3·40	2·35	3·22	4·55	7·69	8·40	14·19	16·65	18·83	19·72	23·87	29·44	34·36
3·50	2·42	3·31	4·66	7·91	8·64	14·60	17·15	19·39	20·30	24·52	30·30	35·36
3·60	2·47	3·41	4·79	8·15	8·89	15·02	17·63	19·87	20·88	25·22	31·16	36·37
3·80	2·60	3·60	5·04	8·60	9·33	15·84	18·57	20·97	22·05	26·63	32·58	38·29
4·00	2·72	3·79	5·31	8·99	9·83	16·68	19·54	22·07	23·22	28·00	34·28	40·31
4·20	2·84	3·98	5·57	9·44	10·31	17·51	20·48	23·17	24·38	29·38	36·00	42·33
4·40	2·98	4·17	5·84	9·89	10·81	18·29	21·43	24·28	25·49	30·77	37·50	44·34
4·50	3·04	4·27	5·94	10·12	11·05	18·70	21·91	24·76	26·08	31·47	38·36	45·26

Projection chart for Berthiot lenses

Slide format 35 × 23·4mm

Image width (m)

Projection distance (m)	25 mm	35 mm	50 mm	80 mm–125 mm zoom	90 mm	150 mm	180 mm	200 mm	210 mm	250 mm	290 mm	360 mm
0·4	0·40	0·285	0·19									
0·6	0·67	0·485	0·33									
0·8	0·95	0·680	0·47									
1·0	1·22	0·885	0·60	0·35 / 0·24	0·32							
1·5	1·90	1·38	0·95	0·55 / 0·37	0·51	0·27						
2·0	2·59	1·88	1·29	0·77 / 0·52	0·70	0·38	0·31	0·27				
2·5	3·27	2·37	1·64	0·98 / 0·66	0·90	0·50	0·40	0·35	0·295	0·23		
3	3·95	2·91	1·99	1·19 / 0·80	1·09	0·61	0·50	0·44	0·46	0·37	0·19	
4	5·32	3·94	2·68	1·60 / 1·08	1·47	0·84	0·70	0·61	0·62	0·50	0·30	0·415
5	6·68	4·98	3·38	2·03 / 1·37	1·86	1·07	0·90	0·79	0·78	0·63	0·42	0·51
6	8·05	6·00	4·07	2·46 / 1·66	2·25	1·30	1·09	0·96	0·94	0·77	0·54	0·60
7	9·41	7·04	4·77	2·87 / 1·94	2·63	1·52	1·28	1·13	1·10	0·90	0·65	0·70
8	10·78	8·07	5·46	3·30 / 2·23	3·02	1·75	1·48	1·30	1·27	1·04	0·77	0·79
9	12·14	9·11	6·16	3·71 / 2·51	3·40	1·98	1·67	1·48	1·43	1·17	0·89	0·88
10	13·51	10·14	6·85	4·15 / 2·81	3·80	2·21	1·87	1·65	1·59	1·31	1·00	
12	16·24	12·21	8·24	4·99 / 3·38	4·56	2·67	2·26	1·99	1·91	1·58	1·12	1·07
14	18·97	14·27	9·63	5·83 / 3·94	5·33	3·12	2·65	2·34	2·24	1·85	1·35	1·26
16	21·70	16·34	11·02	6·67 / 4·51	6·11	3·58	3·04	2·68	2·56	2·12	1·58	1·44
18	24·43	18·40	12·41	7·52 / 5·09	6·88	4·03	3·43	3·03	2·89	2·39	1·82	1·64
20	27·16	20·47	13·80	8·36 / 5·66	7·65	4·49	3·82	3·37	3·21	2·66	2·05	1·82
22	29·89	22·54	15·19	9·19 / 6·22	8·42	4·95	4·21	3·72	3·54	2·93	2·28	2·01
24	32·62	24·60	16·58	10·04 / 6·79	9·19	5·40	4·60	4·06	3·86	3·19	2·52	2·20

Projection chart for Berthiot lenses

Slide format 37·5 × 37·5 mm

Image width (m)

Projection distance (m)	25 mm	35 mm	50 mm	80 mm–125 mm zoom		90 mm	150 mm	180 mm	200 mm	210 mm	250 mm	290 mm	360 mm
0·4	0·43	0·305	0·20										
0·6	0·72	0·52	0·35										
0·8	1·02	0·73	0·50										
1·0	1·31	0·95	0·64	0·37	0·26	0·34							
1·5	2·04	1·48	1·02	0·59	0·40	0·54	0·29						
2·0	2·77	2·01	1·38	0·82	0·56	0·75	0·41	0·33	0·29				
2·5	3·50	2·54	1·76	1·05	0·71	0·96	0·54	0·43	0·37	0·32	0·25		
3	4·23	3·12	2·13	1·27	0·86	1·17	0·65	0·54	0·47	0·49	0·40	0·20	
4	5·70	4·22	2·77	1·71	1·16	1·58	0·90	0·75	0·65	0·66	0·535	0·32	
5	7·16	5·34	3·62	2·17	1·47	1·99	1·15	0·96	0·85	0·84	0·675	0·45	0·45
6	8·62	6·43	4·36	2·64	1·78	2·40	1·39	1·17	1·03	1·00	0·825	0·58	0·55
7	10·08	7·54	5·11	3·07	2·08	2·82	1·63	1·37	1·21	1·18	0·965	0·70	0·64
8	11·55	8·65	5·85	3·54	2·39	3·23	1·87	1·59	1·39	1·36	1·11	0·825	0·75
9	13·00	9·76	6·60	3·97	2·69	3·65	2·12	1·79	1·59	1·53	1·25	0·95	0·85
10	14·47	10·86	7·34	4·45	3·01	4·06	2·37	2·00	1·77	1·70	1·40	1·07	0·94
12	17·40	13·08	8·83	5·35	3·62	4·90	2·86	2·42	2·13	2·05	1·69	1·20	1·15
14	20·32	15·29	10·32	6·25	4·22	5·71	3·34	2·84	2·51	2·40	1·98	1·45	1·35
16	23·25	17·51	11·81	7·15	4·83	6·54	3·84	3·26	2·87	2·74	2·27	1·69	1·54
18	26·17	19·71	13·30	8·06	5·45	7·37	4·32	3·67	3·25	3·10	2·56	1·95	1·76
20	29·10	21·93	14·79	8·96	6·06	8·19	4·81	4·09	3·61	3·44	2·85	2·20	1·95
22	32·02	24·15	16·27	9·85	6·66	9·02	5·30	4·51	3·99	3·79	3·14	2·44	2·15
24	34·95	26·36	17·76	10·76	7·27	9·85	5·79	4·93	4·35	4·14	3·42	2·70	2·36

Appendix 2

Projection Chart for Kodak lenses (used with S-AV 2000 projector)

Slide format (mm)	24 × 36							24 × 36		40 × 40					
Focal length (mm)	35	60	85	100	150	180	250	70–120 mm		60	85	100	150	180	250
Distance (ft/in)	*Picture width (ft/in)*														
1 6	1 7	0 10	0 7	0 5				0 8	0 4	0 11	0 7	0 6			
2 0	2 1	1 2	0 9	0 7				0 11	0 6	1 3	0 10	0 8			
2 6	2 6	1 5	0 12	0 9				1 2	0 8	1 7	1 1	0 10			
3 0	3 0	1 9	1 2	0 12		0 7		1 6	0 10	1 11	1 3	1 1		0 8	
4 0	4 0	2 4	1 7	1 4	0 7	0 10		2 0	1 1	2 6	1 9	1 5	0 7	0 11	
5 0	4 11	2 11	1 12	1 8	0 10	1 0		2 6	1 5	3 2	2 2	1 10	0 11	1 1	1 1
6 0		3 6	2 5	2 0	1 1	1 1		3 0	1 9	3 9	2 7	2 2	1 2	1 6	1 4
8 0		4 8	3 3	2 9	1 3	1 5	0 12	4 0	2 4	5 1	3 6	2 11	1 5	1 11	1 8
10 0		5 10	4 1	3 5	1 9	1 9	1 3	5 0	2 11	6 4	4 5	3 9	1 11	2 4	2 1
12 0		6 12	4 10	4 1	2 3	2 2	1 6	6 0	3 6	7 7	5 3	4 6	2 5	2 9	2 4
14 0		8 2	5 8	4 10	2 9	2 7	1 10	7 0	4 1	8 10	6 2	5 3	2 11	3 3	2 7
16 0		9 4	6 6	5 6	3 2	2 11	2 1	8 0	4 9	10 2	7 1	6 0	3 6	3 7	2 11
18 0		10 6	7 4	6 3	3 8	3 4	2 5	9 0	5 4	11 5	8 0	6 9	4 0	4 1	3 8
20 0		11 8	8 2	6 11	4 7	3 9	2 8	10 0	5 11	12 8	8 10	7 6	5 0	5 1	4 6
25 0		14 7	10 2	8 8	5 9	4 8	3 5	12 6	7 5	15 10	11 1	9 5	6 3	6 2	5 3
30 0		17 6	12 3	10 5	6 12	5 8	4 1		8 11	19 0	13 3	11 4	7 7	7 2	6 0
35 0		20 5	14 3	12 2	8 2	6 7	4 10			22 2	15 6	13 2	8 10	8 2	7 1
40 0		23 4	16 4	13 11	9 4	7 7	5 7			25 4	17 9	15 1	10 2	10 4	9 1
50 0			20 5	17 5	11 8	9 6	6 12				22 2	18 11	12 9	12 5	12 3
60 0			24 6	20 11	14 1	11 5	8 5				26 7	22 8	15 3	16 8	15 4
80 0					18 10	15 4	11 3						20 5	20 10	
100 0					23 6	19 2	14 1						25 7		

215

Projection chart for Navitar lenses

Slide format 1·34 × 902 in

Projected image		Projection distances (ft)					
width	height	1″	1½″	2″	70 mm–115 mm ZOOM 2¾″–5″	7″	9″
1 2	8	1 2	1 6	1 10	2 7– 4 5	6 4	8 4
1 6	1	1 7	2 1	2 9	3 7– 6 2	8 9	11 6
2	1 4	1 11	2 8	3 5	4 9– 8 3	11 7	15 1
3	2	2 8	3 9	4 9	6 8–11 10	16 9	21 7
3 4	2 2½	2 10	4 2	5 6	7 1–12 11	18 3	23 6
4 2	2 6½	3 6	5 2	6 6	9 2–16 4	23	29 7
5	3 4	4 2	5 11	7 11	10 10–19 3	27 2	34 10
6 10	3 11	4 9	6 10	8 10	12 4–22 3	31 4	40 3
7	4 8½	5 8	8 1	10 9	15 –26 8	37 6	48 3
8	5 4½	6 4	9 3	12 2	16 10–30 5	42 9	54 11
9	6	7 2	10 5	13 8	18 11–34 1	48	61 8
10	6 8½	7 10	11 7	15 2	21 1–37 10	53 2	68 4
12	8	9 4	13 9	18 2	25 –45 3	63 7	81 9
14	9 4½	10 10	15 11	21 2	29 1–52 8	73 10	95 1
16	10 8½	12 4	18 3	24 1	33 3–60 1	84 5	108 6
18	12 3	13 10	20 6	27 1	37 3–67 6	94 8	121 9
20	13 5	15 3	22 8	30 1	41 4– 75	105 2	135 3

All measurements in feet and inches

Projection chart for Buhl lenses

Slide format 1·34 × ·902 in

Projected image		Projection distances (f)				
width	1·0	1·4	1·4 HI-SPD	2·0	2·0 HI-SPD	height
18	20	25	26	32·5	34	12
20	21·5	27	28	35	37	13
22	23	29	30	38	39·5	14·5
24	24·5	31·5	32·5	40·5	42·5	16
26	26	33·5	34·5	43	45	17
28	27·5	35·5	37	46	48	18·5
30	29	37·5	39	49	51	20
32	30	39·5	41	51·5	54	21
34	31·5	42	43	54	56·5	22·5
36	33	44	45·5	57	59·5	24
38	34·5	46	47	59·5	62	25
40	36	48	50	62	65	26·5
42	37·5	50	52	65	68	28
44	39	52	54	68	71	29
46	40·5	54	56	70·5	73·5	30·5
48	42	56	58·5	73·5	76·5	32
50	43	58	60·5	76	79	33
52	44·5	60	63	79	82	34·5
54	46	62·5	65	82	85	36
56	47	64·5	67	84·5	88	37
58	48	66·5	69	87	90·5	38·5
60	49·5	68·5	71·5	90	93·5	40
66	55	75	78	98·5	102	44
72	59·5	81	85	106·5	111	48·5
78	63·5	87	91	115	119·5	52·5
84	68	93·5	97·5	123	128	56·5
90	72·5	99·5	104	131	136·5	60·5
96	76·5	106	111	139·5	145	64·5
102	81	112	117·5	147·5	153·5	68·5
108	85·5	118	124	155·5	162	72·5
114	90	124·5	130	164	170·5	76·5
120	94	130·5	137	172	179	80·5
126	98·5	137	143·5	180·5	187·5	84·5
132	102·5	143	150	188·5	196	88·5
138	107	149	156·5	197	204·5	92·5
144	111·5	155·5	163	205	213	96·5

All measurements are in inches. . . .

(1) All measurements are made from the film plane.
(2) Because focal length tolerances and lens construction vary from model to model, this chart assumes that projection distances can be varied by at least, (plus or minus), 3%.
(3) When more precise measurements or very exact screen width/distance relationships are required, Buhl is prepared to specially select lenses.

Projection chart for Buhl lenses

Projected image	Projected distances	
width, height	2·0	2·0 Hi Spd
18	29·2	30·5
20	31·6	33·1
22	34·1	35·6
24	36·5	38·2
26	39·0	40·7
28	41·5	43·3
30	43·9	45·8
32	46·4	48·4
34	48·8	50·9
36	52·3	53·5
38	53·7	56·0
40	56·2	58·6
42	58·7	61·1
44	61·1	63·7
46	63·6	66·2
48	66·0	68·8
50	68·5	71·3
52	71·0	73·9
54	73·4	76·4
56	75·9	79·0
58	78·3	81·5
60	80·8	84·1
66	88·2	91·7
72	95·5	99·4
78	102·9	107·1
84	110·3	114·7
90	117·7	122·4
96	125·0	130·0
102	132·4	137·7
108	139·8	145·3
114	147·2	153·0
120	154·5	160·6
126	161·9	168·3
132	169·3	176·0
138	176·7	183·6
144	184·0	191·3

(1) All measurements are made from the film plane.
(2) Because focal length tolerances and lens construction vary from model to model, this chart assumes that projection distances can be varied by at least, (plus or minus), 3 per cent.
(3) When more precise measurements or very exact screen width/distance relationships are required, Buhl is prepared to specially select lenses.

Projection distance table for Kodak Ektagraphic projectors

Projection distances (m)										Screen image dimensions (m)				
1·4 *35*	2 *50*	2½ *65*	3 *75*	4 *100*	5 *125*	7 *180*	9 *230*	11 *280*	4 to 6 (zoom) *100–150*	*135–35* mm	*126*	*Super-slide*	*Single-frame filmstrip*	*110*
2 *0·6*	3 *0·8*	3½ *1·1*	4 *1·3*	5½ *1·7*	7 *2·1*	10 *3·1*	12½ *3·9*	15½ *4·8*	5½–8½ *1·7–2·5*	13½ × 20 *0·34 × 0·51*	15½ sq *0·40*	22 sq *0·57*	10 × 13 *0·25 × 0·34*	7 × 9 *0·18 × 0·24*
3 *0·8*	4 *1·2*	5 *1·6*	6 *1·8*	8 *2·4*	10 *3·0*	14 *4·4*	18 *5·6*	22½ *6·8*	8–12 *2·4–3·6*	20 × 30 *0·51 × 0·76*	23 sq *0·59*	33½ sq *0·84*	15 × 19½ *0·38 × 0·50*	10½ × 14 *0·27 × 0·35*
3½ *1·1*	5½ *1·6*	6½ *2·1*	8 *2·4*	10½ *3·2*	13 *3·9*	18½ *5·7*	24 *7·3*	29 *8·8*	10½–16 *3·2–4·7*	27 × 40 *0·68 × 1·01*	31 sq *0·78*	44½ sq *1·12*	20 × 26½ *0·51 × 0·66*	14 × 18½ *0·35 × 0·47*
4½ *1·4*	6½ *2*	8 *2·5*	10 *2·9*	13 *3·9*	16½ *4·9*	23 *7·1*	29½ *9·0*	36 *11·0*	13–19½ *4–5·9*	33½ × 50 *0·85 × 1·27*	38½ sq *0·98*	55½ sq *1·41*	25 × 33 *0·63 × 0·84*	17½ × 23 *0·45 × 0·59*
5½ *1·6*	8 *2·3*	9½ *3·0*	11½ *3·5*	15½ *4·6*	19½ *5·8*	27 *8·4*	35 *10·7*	42½ *13·0*	15½–23½ *4·8–7·0*	40 × 60 *1·02 × 1·52*	46½ sq *1·18*	66½ sq *1·68*	30 × 39½ *0·76 × 1*	21 × 27½ *0·53 × 0·70*
6½ *1·9*	9½ *2·8*	11½ *3·6*	14 *4·2*	18½ *5·6*	23 *6·9*	32½ *10·0*	41½ *12·8*	51 *15·6*	18½–28 *5·5–8·3*	48 × 72 *1·23 × 1·83*	56 sq *1·42*	80 sq *2·03*	36 × 47½ *0·91 × 1·20*	25½ × 33½ *0·64 × 0·85*
8½ *2·6*	12 *3·7*	15½ *4·8*	18½ *5·5*	24½ *7·3*	30½ *9·2*	43 *13·2*	55 *16·9*	67 *20·5*	24½–36½ *7·5–11·0*	64½ × 96 *1·63 × 2·44*	74½ sq *1·89*	106½ sq *2·71*	47½ × 63 *1·21 × 1·61*	33½ × 44½ *0·86 × 1·13*
10½ *3·2*	15 *4·6*	19 *5·9*	23 *6·8*	30½ *9·1*	38 *11·4*	53 *16·4*	68½ *21·0*	83½ *25·5*	30½–45½ *9·3–13·7*	80½ × 120 *2·04 × 3·05*	93 sq *2·36*	133½ sq *3·39*	60 × 79 *1·52 × 2·01*	42 × 55½ *1·07 × 1·41*
12½ *3·8*	18 *5·5*	22½ *7·1*	27 *8·2*	36½ *10·9*	45½ *13·6*	63½ *19·6*	81½ *25·1*	100 *30·5*	36½–54½ *11·2–16·4*	96½ × 144 *2·45 × 3·66*	111½ sq *2·84*	160 sq *4·07*	71½ × 94½ *1·82 × 2·41*	50½ × 66½ *1·28 × 1·69*

Projection distances are approximate and are measured from projector gate to screen.

Appendix 3

Superslide coverage

Superslides using 46mm film frequently need to be rear projected in the same way as any other format. Unfortunately, the extra area of the slide can rarely be projected by short focal length lenses that are designed for the standard 35mm format. As a guide I have listed nine popular lenses with an indication of how well they can cover superslides.

Lens	*Slide aperture*		
	37·5 × 37·5 mm	35 × 35 mm	34 × 34 mm
Berthiot 25 mm	A	A	A
Berthiot 35 mm	A	B	B
Berthiot 50 mm	B	B	B
Berthiot 80/125 mm zoom	C	C	C
Buhl Cat No. 475–800	A	A	A
Buhl Cat No. 891–800	A	A	A
Kodak 35 mm (with special condenser)	A	B	B
Kodak 70–120 mm zoom	C	C	C

A: does not cover at all.
B: covers complete picture area but with slight darkening at corners, satisfactory for most purposes.
C: covers complete picture area with even illumination.

Appendix 4

Projection formats

Description	Aperture width (in)	(mm)	Aperture height (in)	(mm)	Aspect ratio width/height (in, mm)
8 mm Motion Picture	0·17	4·37	0·12	3·28	1·33
Super 8 Motion Picture	0·21	5·36	0·15	4·01	1·33
16 mm Motion Picture	0·38	9·65	0·28	7·21	1·33
16 mm Cinemascope	0·38×2	9·65	0·28	7·21	2·66
126 Insta-Load Slides	0·66	17·0	0·50	12·7	1·34
35 mm Motion Picture	0·82	20·9	0·60	15·2	1·37
35 mm Cinemascope					2·34
35 mm Filmstrip	0·91	23·1	0·69	17·5	1·32
2 × 2 Half Frame	0·90	22·9	0·62	15·9	1·44
2 × 2 Standard 35 mm Double Frame Slide	1·34	34·2	0·90	22·9	1·49
2 × 2 Instamatic	1·04	26·5	1·04	26·5	1·00
2 × 2 Superslides	1·5	38·1	1·5	38·1	1·00
2¼ × 2¼ Slides	2·03	51·6	2·03	51·6	1·00
3¼ × 4 Lantern Slide	3·00	76·2	2·25	57·2	1·33
3¼ × 4 Polaroid	3·26	82·8	2·4	61·0	1·36
4 × 5	4·50	114·3	3·50	88·9	1·28
Overhead Projector	10·0	254	10·0	254	1·00
Overhead Projector	9·5	241	7·5	190·5	1·26
Television Projectors					1·33

Aspect ratio

This is a convenient method of relating image width to image height, eg. to find height of a 16 mm image 5 feet wide divide 5 by 1·33 (the aspect ratio) to obtain image height of 3·76 feet.

Appendix 5

Playing time for various tape speeds and reel sizes

All tapes shown are standard 1-1/2 mil except as indicated.

Reel size (in)	Tape length (ft)	Playing time for various tape speeds and tape lengths — Tape speed–(ips)				Dual track time	
		1-7/8 ips	3-3/4 ips	7-1/2 ips	15 ips	3-3/4 ips	7-1/2 ips
3	150	15 min.	7-1/2 min.	3-3/4 min.	1-7/8 min.	15 min.	7-1/2 min.
3	225*	22-1/2 min.	11-1/4 min.	5-5/8 min.	2-15/16 min.	22-1/2 min.	11-1/4 min.
4	300	30 min.	15 min.	7-1/2 min.	3-3/4 min.	30 min.	15 min.
4	450*	45 min.	22-1/2 min.	11-1/4 min.	5-5/8 min.	45 min.	22-1/2 min.
5	600	1 hr.	30 min.	15 min.	7-1/2 min.	1 hr.	30 min.
5	900*	1-1/2 hr.	45 min.	22-1/2 min.	11-1/4 min.	1-1/2 hr.	45 min.
7	1200	2 hr.	1 hr.	30 min.	15 min.	2 hr.	1 hr.
7	1800*	3 hr.	1-1/2 hr.	45 min.	22-1/2 min.	3 hr.	1-1/2 hr.
7	2400**	4 hr.	2 hr.	1 hr.	30 min.	4 hr.	2 hr.
10-1/2	2400	4 hr.	2 hr.	1 hr.	30 min.	4 hr.	2 hr.
10-1/2	3600*	6 hr.	3 hr.	1-1/2 hr.	45 min.	6 hr.	3 hr.
14	4800	8 hr.	4 hr.	2 hr.	1 hr.	8 hr.	4 hr.
14	7200*	12 hr.	6 hr.	3 hr.	1-1/2 hr.	12 hr.	6 hr.

*Long playing tape (1 mil film)
**Double play tape (1/2 mil film)

Appendix 6

Movie film data

For 35mm and 16mm film.

Standard speed (feet per minute)	90	36
Standard frame rate (per second)	24	24
Perforations per frame	8	1 (sound)
	(4 each side)	2 (silent)
Frame sizes		
(camera aperture)	$0 \cdot 631 \times 0 \cdot 868$ in	$0 \cdot 294 \times 0 \cdot 410$ in
(projector aperture)	$0 \cdot 600 \times 0 \cdot 825$ in	$0 \cdot 284 \times 0 \cdot 380$ in
Frames per foot	16	40
Footage for 10 minutes running time	900	360

Appendix 7

World electricity supply guide

Country	Frequency and Tolerance (Hz, %)	Household voltage (V)	Commercial voltage (V)	Industrial voltage (V)	Voltage tolerance (%)
AFGHANISTAN	50	380/220 (A) 220 (L)	380/220 (A)	380/220 (A) (3)	(9)
ALGERIA	50±1·5	220/127 (E) 220 (L) (I)	380/220 (A) 220/127 (A)	10 kV 5·5 kV 6·6 kV 380/220 (A)	±5 and±10
ANGOLA	50	220 (L) (I)	380/220 (A)	380/220 (A) (3)	(9)
ANGUILLA	50	230 (L) (I)	400/230 (A)	400/230 (A) (3)	(9)
ANTIGUA	60	230 (L) (I)	400/230 (A)	400/230 (A) (3)	(9)
ARGENTINA	50±1·0	225 (L) (I) 220 (L) (I)	390/225 (A) 380/220 (A) 220 (L)	13·2 kV 6·88 kV 390/225 (A) 380/220 (A)	±10
AUSTRALIA	50±0·1	415/240 (A) (E) 240 (L)	415/240 (A) 440/250 (A) 440 (N) (6)	22 kV 11 kV 6·6 kV 415/240 (A) 440/250 (A)	±6
AUSTRIA	50±0·1	380/220 (A) (B) 220 (L)	380/220 (A) (B) 220 (L)	20 kV 10 kV 5 kV 380/220 (A)	±5
BAHAMAS	60	240/120 (G) 120 (L)	240/120 (G) 120 (L)	415/240 (A) (3) 208/120 (A)	(9)
BAHRAIN	50 & 60	400/230 (A) 230 (L) 110 (L)	400/230 (A) 380/220 (A) 230 (L) 220/110 (K)	11 kV 400/230 (A) 380/220 (A)	±6

Country	Frequency (Hz)				Voltage tolerance
BANGLADESH	50±4	400/230 (A) 230 (L)	11 kV 400/230 (A)	11 kV 400/230 (A)	±5
BARBADOS	50±0·4	230/115 (G) (K) 200/115 (A) (E)	230/115 (G) (K) 200/115 (A)	11 kV 3·3 kV 230/115 (G) 200/115 (A)	±6
BELGIUM	50±3	380/220 (A) 220/127 (A) 220 (F)	380/220 (A) 220/127 (A) 220 (F)	15 kV 6 kV 380/220 (A) 220/127 (A) 220 (F)	±5 (day) ±10 (night)
BELIZE	60±0·1	220/110 (K)	220/110 (K)	440/220 (K) (3)	(9)
BERMUDA	60±0·1	240/120 (K) 208/120 (A)	240/120 (K) 208/120 (A)	4·16/2·4 kV 208/120 (A) 240/120 (K)	±5
BOLIVIA	50±1	230/115 (H)	230/115 (H)	230/115 (H) (3)	±5
BOTSWANA	50	220 (L) (1)	380/220 (A)	380/220 (A) (3)	(9)
BRAZIL	60	220 (L) (1) 127 (L) (1)	380/220 (A) 220/127 (A)	13·8 kV 11·2 kV 380/220 (A) 220/127 (A)	(9)
BULGARIA	50±0·1	380/220 (A) 220 (L)	380/220 (A) 220 (L)	20 kV 15 kV 380/220 (A)	±5
BURMA	50	230 (L) (1)	400/230 (A) 230 (L)	11 kV 6·6 kV 400/230 (A)	(9)
CAMBODIA	50	208/120 (A) 120 (L)	380/220 (A) 208/120 (A)	380/220 (A) (3) 208/120 (A)	(9)
CAMEROON (FR)	50±2	220 (L) (1)	380/220 (A)	15 kV 380/220 (A)	±5

Country	Frequency and Tolerance (Hz, %)	Household voltage (V)	Commercial voltage (V)	Industrial voltage (V)	Voltage tolerance (%)
CANADA	60±0·02	240/120 (K)	600/347 (A) 480 (F) 240 (F) 240/120 (K) 208/120 (A)	12·5/7·2 kV 600/347 (A) 208/120 (A) 600 (F) 480 (F) 240 (F)	±4 8·3
CAYMAN ISLANDS	60±0·1	240/120 (K)	240/120 (K) (G)	480/240 (G) 480/227 (A) 240/120 (G) 208/120 (A)	±10
CHAD	50	220 (L) (1)	220 (L) (1)	380/220 (A) (3)	(9)
CHILE	50	220 (L) (1)	380/220 (A) (1)	380/220 (A) (3)	(9)
CHINA (PR)	50	220 (L) (1)	380/220 (A)	380/220 (A) (3)	±7
COLOMBIA	60±1	240/120 (G) 120 (L)	240/120 (G) 120 (L)	13·2 kV 240/120 (G)	±10
COSTA RICA	60	120 (L) (1)	240/120 (K) 120 (L) (1)	240/120 (G) (3)	(9)
CYPRUS	50±2·5	240 (L) (1)	240 (L) (1)	11 kV 115/240 (A)	±6
CZECHOSLOVAKIA	50±0·1	380/220 (A) 220 (L)	380/220 (A) 220 (L)	22 kV 15 kV 6 kV 3 kV 380/220 (A)	±10
DENMARK	50±0·4	380/220 (A) 220 (L)	380/220 (A) 220 (L)	30 kV 10 kV 380/220 (A)	±10
DAHOMEY	50±1	380/220 (A) 220 (L)	380/220 (A) 220 (L)	15 kV 380/220 (A)	±10
DOMINICA	50	230 (L) (1)	400/230 (A)	400/230 (A) (3)	(9)

Country	Frequency				Tolerance
DOMINICAN REPUBLIC	60	110 (L) (1)	220/110 (K) (1) 110 (L)	220/110 (G) (3)	(9)
ECUADOR	60	127 (L) (1) 120 (L) (1) 110 (L)	240/120 (K) 208/120 (A) 220/127 (A) 220/110 (K)	240/120 (K) 208/120 (A) 220/127 (A) 220/110 (K)	(9)
EGYPT (AR)	50±1	380/220 (A) 220 (L)	380/220 (A) 220 (L)	11 kV 6·6 kV 380/220 (A)	±10
EL SALVADOR	60±1	240/120 (K)	240/120 (K) (G)	14·4 kV 2·4 kV 240/210 (G)	±5
ETHIOPA	50	220 (L) (1)	380/220 (A)	380/220 (A) (3)	(9)
FALKLAND ISLANDS	50±3	230 (L) (1)	415/230 (A)	415/230 (A) (3)	±2·5
FIJI ISLANDS	50±1	415/240 (A) 240 (L)	415/240 (A) 240 (L)	11 kV 415/240 (A)	(9)
FINLAND	50±0·1	220 (L) (1)	380/220 (A)	660/380 (A) 500 (B) 380/220 (A) (D)	±10
FRANCE	50±1	380/220 (A) 220 (L) 220/127 (A) 127 (L)	380/220 (A) 380/220 (D) 380 (B)	20 kV 15 kV 380 (B) 380/220 (A) (D)	±10
GAMBIA	50	230 (A) (1)	230 (A) (1)	400/230 (A) (3)	±5(1)
GERMANY (FR)	50±0·3	380/220 (A) 220 (L)	380/220 (A) 220 (L)	20 kV 10 kV 380/220 (A)	±10
GERMANY (DDR)	50±0·3	380/220 (A) 220 (L) 220/127 (A) 127 (L)	380/220 (A) 220 (L)	10 kV 6 kV 660/380 (A) 380/220 (A)	±5
GHANA	50±5	250 (L) (1)	250 (L) (1)	440/250 (A) (3)	±10
GIBRALTAR	50±1	415/240 (A)	415/240 (A)	415/240 (A) (3)	±6

227

Country	Frequency and Tolerance (Hz, %)	Household Voltage (V)	Commercial Voltage (V)	Industrial Voltage (V)	Voltage tolerance (%)
GREECE	50±1	220 (L) (1)	6·6 kV 380/220 (A)	22 kV 20 kV 15 kV 6·6 kV 380/220 (A)	±5
GRENADA	50	230 (L) (1)	400/230 (A)	400/230 (A) (3)	(9)
GUADELOUPE	50 & 60	220 (L) (1)	380/220 (A)	20 kV 380/220 (A)	(9)
GUAM (Mariana Islands)	60+1 −0·08	240/120 (K) 208/120 (A) 240 (L) 120 (L)	240/120 (K) 208/120 (A)	13·8 kV 4·0 kV 480/227 (A) 480 (D) 240/120 (H) 208/120 (A)	+8, −10
GUATEMALA	60±1·7	240/120 (K)	240/120 (K)	13·8 kV 240/120 (G)	±10
HAITI	60	230 (L) (1) 220 (L) (1) 115 (L)	380/220 (A) 230/115 (K) 220 (L)	380/220 (A) 230/115 (G)	(9)
HONDURAS	60	110 (L)	220/110 (K) 110 (L)	220/110 (K) (3)	(9)
HONG KONG (and Kowloon)	50±2	346/200 (A) 200 (L) (1)	11 kV 346/200 (A) 380/220 (A) 200 (L)	11 kV 346/200 (A) 380/220 (A) (3)	±6
HUNGARY	50±2	380/220 (A) 220 (L)	380/220 (A) 220 (L)	20 kV 10 kV 380/220 (A)	+5, −10
ICELAND	50±0·1	380/220 (A) 220 (L)	380/220 (A) 220 (L)	380/220 (A) (3)	(9)

Country	Frequency				Tolerance
INDIA (4)					
Bombay	50±1	440/250 (A) 230 (L)	440/250 (A) 230 (L)	11 kV 440/250 (A)	±4
New Delhi	50±3	400/230 (A) 230 (L)	400/230 (A) 230 (L)	11 kV 400/230 (A)	±6
Ramakrishnapuram (2)	50±3 25 d.c.	400/230 (A) 230 (L) 460/230 (P)	400/230 (A) 230 (L) 460/230 (P)	22 kV & 11 kV (9) (9)	±6
INDONESIA	50+1 −2	220/127 (A)	380/220 (A) 220/127 (A)	380/220 (A) (3)	±5
IRAN	50±5	220 (L) (1)	380/220 (A)	20 kV 11 kV 400/231 (A) 380/220 (A)	±15
IRAQ	50	220 (L) (1)	380/220 (A)	11 kV 6·6 kV 3 kV 380/220 (A)	±5
IRELAND NORTHERN	50±0·4	230 (L) (1) 220 (L) (1)	400/230 (A) 380/220 (A)	400/230 (A) (3) 380/220 (A)	±6
IRELAND REPUBLIC OF	50	220 (L) (1)	380/220 (A)	10 kV 380/220 (A)	(9)
ISRAEL	50±0·2	400/230 (A) 230 (L)	400/230 (A) 230 (L)	22 kV 12·6 kV 6·3 kV 400/230 (A)	±6
ITALY	50±0·4	380/220 (A) 220/127 (E) 220 (L)	380/220 (A) 220/127 (E)	20 kV 15 kV 10 kV 380/220 (A) 220 (C)	±5 (urban) ±10 (rural)
IVORY COAST	50	220 (L) (1)	380/220 (A)	380/220 (A) (3)	(9)
JAMAICA	50±1	220/110 (G) (K)	220/110 (G) (K)	4/2·3 kV 220/110 (G)	±6
JAPAN (EAST) (4)	50±0·2 (5)	200/100 (K) 100 (L)	200/100 (H) (K)	6·6 kV 200/100 (H) 200 (G) (J)	±10

Country	Frequency and Tolerance (Hz, %)	Household voltage (V)	Commercial voltage (V)	Industrial voltage (V)	Voltage tolerance (%)
JAPAN (WEST) (4)	60±0·1 (5)	210/105 (K) 200/100 (K) 100 (L)	210/105 (H) (K) 200/100 (K) 100 (L)	22 kV 6·6 kV 210/105 (H) 200/100 (H)	±10
JORDAN	50	380/220 (A) 220 (L)	380/220 (A)	380/220 (A) (3)	(9)
KENYA	50	240 (L) (1)	415/240 (A)	415/240 (A) (3)	(9)
KOREA, NORTH	60+0 −5	220 (L)	380/220 (A)	380/220 (A)	+6·8 (10) −13·6
KOREA, SOUTH	60	100 (L)	200/100 (K)	(9)	(9)
KUWAIT	50	240 (L) (1)	415/240 (A)	415/240 (A) (3)	(9)
LAOS	50±8	380/220 (A)	380/220 (A)	380/220 (A) (3)	±6
LEBANON	50	220 (L) (1) 110 (L) (1)	380/220 (A) 220 (L) 190/110 A 110 (L)	380/220 (A) (3) 190/110 (A)	(9)
LESOTHO	50	220 (L) (1)	380/220 (A)	380/220 (A) (3)	(9)
LIBERIA	60±3·3	240/120 (K)	240/120 (K)	12·5/7·2 kV 416/240 (B) 240/120 (K) 208/120 (D)	±1·7
LIBYA	50	230 (L) (1) 127 (L) (1)	400/230 (A) 220/127 (A) 230 (L) 127 (L)	400/230 (A) (3) 220/127 (A)	(9)
LUXEMBOURG	50±0·5	380/220 (A) 220/127 (A) 208/120 (A)	380/220 (A) 220/127 (A) 208/120 (A)	20 kV 15 kV 5 kV	±5 and ±10

Country	Frequency	Voltage (L)	Voltage (A)	Higher voltages	Tolerance
MALAGASY REPUBLIC	50±2	220 (L) (1) 127 (L) (1)	380/220 (A) 220/127 (A)	5 kV 380/220 (A) 220/127 (A)	±3
MALAWI	50	230 (L) (1)	400/230 (A)	400/230 (A) (3)	(9)
MALAYSIA	50±1·0	240 (L) (1)	415/240 (A)	415/240 (A) (3)	+5 +10
MALI	50	220 (L) (1) 127 (L) (1)	380/220 (A) 220/127 (A)	380/220 (A) (3) 220/127 (A)	(9)
MALTA	50±1	240 (L) (1)	415/240 (A)	11 kV 6·6 kV 3·3 kV 415/240 (A)	(9)
MARTINIQUE	50	127 (L) (1)	220/127 (A) 127 (L)	220/127 (A) (3)	(9)
MAURITIUS	50±1·0	230 (L) (1)	400/230 (A)	400/230 (A) (3)	±6
MEXICO	60±0·2	220/127 (A) 220 (L) 120 (M)	220/127 (A) 220 (L) 120 (M)	13·8 kV 13·2 kV 480/277 (A) 220/127 (B)	±6
MONACO	50	380/220 (A) 220 (L) 220/127 (A) 127 (L)	380/220 (A) 220 (L)	380/220 (A) (3)	(9)
MONTSERRAT	60	230 (L) (1)	400/230 (A)	400/230 (A) (3)	(9)
MUSCAT & OMAN	50	240 (L) (1)	415/240 (A) 240 (L)	415/240 (A) (3)	(9)
MOROCCO	50	220/127 (A) 200/115 (A)	380/220 (A)	380/220 (A) (3)	(9)
NEPAL	50±1	220 (L) (1)	400/220 (A) 220 (L)	11 kV 400/220 (A)	±10
NETHERLANDS	50±0·4	380/220 (A) 220 (E) (L)	380/220 (A)	10 kV 3 kV 380/220 (A)	±6

Country	Frequency and Tolerance (Hz, %)	Household Voltage (V)	Commercial Voltage (V)	Industrial Voltage (V)	Voltage tolerance (%)
NETHERLANDS ANTILLES	50 & 60	220 (L) (1) 127 (L) (1) 120 (L) (1) 115 (L) (1)	380/220 (A) 230/115 (K) 220/127 (A) 208/120 (A)	380/220 (A) (3) 230/115 (G) 220/127 (A) 208/120 (A)	(9)
NEW GUINEA	50±2	240 (L) (1)	415/240 (A) 240 (L)	22 kV 11 kV 415/240 (A)	±5
NEW ZEALAND	50±1·5	400/230 (A) (E) 230 (L) 240 (L)	415/240 (A) (E) 400/230 (A) (E) 230 (L) 240 (L)	11 kV 400/230 (A) 415/240 (A) 440 (N) (6)	±5
NICARAGUA	60	240/120 (G) (K)	240/120 (G) (K)	13·2 kV 7·6 kV 240/120 (G)	(9)
NIGERIA	50±1	230 (L) (1) 220 (L) (1)	400/230 (A) 380/220 (A)	15 kV 11 kV 400/230 (A) 380/220 (A)	±5
NIGER	50±1	220 (L) (1)	380/220 (A)	15 kV 380/220 (A)	±2·5
NORWAY	50±0·2	230 (B)	380/220 (A) 230 (B)	20 kV 10 kV 5 kV 380/220 (A) 230 (B)	±10
PAKISTAN	50	230 (L) (1)	400/230 (A)	400/230 (A) (3)	(9)
PANAMA	60±0·17	240/120 (K)	480/277 (A) 240/120 (K)	12 kV 480/277 (A) 208/120 (A)	±5

Country	Frequency				Tolerance (%)
PARAGUAY	50	220 (L) (1)	440/220 (K) 380/220 (A)	440/220 (G) (3) 380/220 (A)	(9)
PERU	60	225 (B) (M)	225 (B) (M)	10 kV 6 kV 225 (B)	(9)
PHILIPPINES	60±0·16	220/110 (K)	13·8 kV 4·16 kV 2·4 kV 220/110 (H)	13·8 kV 4·16 kV 2·4 kV 440 V (B) 220/110 (H)	±5
Manila (*Metropolitan area*)	60±5	240/120 (H) (K) 240/120 (H)	240/120 (H) (K) 240/120 (H)	20 kV 6·24 kV 3·6 kV 240/120 (H)	±5
POLAND	50±1	220 (L) (1)	380/220 (A)	15 kV 6 kV 380/220 (A)	±5
PORTUGAL	50±1	380/220 (A) 220 (L)	15 kV 5 kV 380/220 (A) 220 (L)	15 kV 5 kV 380/220 (A)	±5
PUERTO RICO	60±10	240/120 (L)	480 (F) 240/120 (L)	8·32 kV 4·16 kV 480 (F)	±10
QATAR	50	240 (L) (1)	415/240 (A) 240 (L)	415/240 (A) (3)	±6
RHODESIA	50±2·5	225 (L) (1)	390/225 (A)	11 kV 390/225 (A)	±6·6
ROMANIA	50±1	220 (L) (1)	380/220 (L)	20 kV 10 kV 6 kV 380/220 (A)	±5
RWANDA	50±1	220 (L) (1)	380/220 (A)	15 kV 6·6 kV 380/220 (A)	±5

Country	Frequency and Tolerance (Hz, %)	Household voltage (V)	Commercial voltage (V)	Industrial voltage (V)	Voltage tolerance (%)
SABAH	50±0·5	240 (L) (1)	415/240 (A)	415/240 (A) (3)	±6
SAUDI ARABIA	60±0·5 50±0·5	220/127 (A) 127 (L)	380/220 (A) 220/127 (A) 127 (L)	380/220 (A) (3) 220/127 (A)	±5
SENEGAL	50	127 (L) (1)	220/127 (A) 127 (L)	220/127 (A) (3)	(9)
SIERRA LEONE	50	230 (L) (1)	400/230 (A) 230 (L)	11 kV 400/230 (A)	(9)
SINGAPORE	50±0·5	400/230 (A) 230 (L)	6·6 kV 400/230 (A)	22 kV 6·6 kV 400/230 (A)	±3
SOMALI REPUBLIC	50	230 (L) 220 (L) 110 (L) (1)	440/220 (K) 220/110 (K) 230 (L)	440/220 (G) (3) 220/110 (G)	(9)
SOUTH AFRICA	50±2·5 25 (8)	433/250 (A) (7) 400/230 (A) (7) 380/220 (A) 220 (L)	11 kV 6·6 kV 3·3 kV 433/250 (A) (7) 400/230 (A) (7) 380/220 (A)	11 kV 6·6 kV 3·3 kV 500 (B) 380/220 (A)	±5
SPAIN	50±3	380/220 (A) (E) 220 (L) 220/127 (A) (E) 127 (L)	380/220 (A) 220/127 (A)	15 kV 11 kV 380/220 (A)	±7
SRI LANKA (Ceylon)	50±2	230 (L) (1)	400/230 (A) 230 (L)	11 kV 400/230 (A)	±6
ST. KITTS & NEVIS	60	230 (L) (1)	400/230 (A)	400/230 (A) (3)	(9)
ST. LUCIA	50	240 (L) (1)	415/240 (A)	11 kV 415/240 (A)	(9)

Country	Frequency				Note
ST. VINCENT	50	230 (L) (1)	400/230 (A)	3·3 kV 400/230 (A)	(9)
SUDAN	50	240 (L) (1)	415/240 (A) 240 (L)	415/240 (A) (3)	(9)
SURINAM	50 & 60	115 (L) 127 (L) (1)	230/115 (K) 220/127 (A) 220/110 (K)	230/115 (G) (3) 220/127 (A) 220/110 (G)	(9)
SWAZILAND	50±2·5	230 (L) (1)	400/230 (A) 230 (L)	11 kV 400/230 (A)	±6
SWEDEN	50±0·2	380/220 (A) 220 (L)	380/220 (A) 220 (L)	20 kV 10 kV 6 kV 380/220 (A)	±10
SWITZERLAND	50±0·5	380/220 /A 220 (L)	380/220 (A) 220 (L)	16 kV 11 kV 6 kV 380/220 (A)	±10
SYRIA	50	220 (L) (1) 115 (L) (1)	380/220 (A) 220 (L) 200/115 (A) 115 (L)	380/220 (A) (3) 200/115 (A)	(9)
TAIWAN (FORMOSA)	60±4	380/220 (A) 220 (L) 220/110 (K) 110 (L)	380/220 (A) 220/110 (H)	22·8 kV 11·4 kV 380/220 (A) 220 (H)	±5 and ±10
TANZANIA	50	400/230 (A)	400/230 (A)	11 kV 400/230 (A)	(9)
THAILAND	50	220 (L) (1)	380/220 (A) 220 (L)	380/220 (A) (3)	(9)
TOGO	50	220 (L) (1)	380/220 (A)	20 kV 5·5 kV 380/220 (A)	(9)
TONGA	50	415/240 (A) 240 (L) 110 (L)	415/240 (A) 240 (L) 110 (L)	11 kV 6·6 kV 415/240 (A)	(9)

Country	Frequency and Tolerance (Hz, %)	Household voltage (V)	Commercial voltage (V)	Industrial voltage (V)	Voltage tolerance (%)
TRINIDAD & TOBAGO	60±0·5	230/115 (K)	400/230 (A) 230/115 (G)	12 kV 400/230 (A)	±6
TUNISIA	50±2	380/220 (A) 220 (L)	380/220 (A) 220 (L)	15 kV 10 kV 380/220 (A)	±10
TURKEY	50±2	220 (L) (1)	380/220 (A)	15 kV 6·3 kV 380/220 (A)	±10
UGANDA	50±0·1	240 (L) (1)	415/240 (A)	11 kV 415/240 (A)	±4·5
UNITED ARAB EMIRATES Dubai	50±0·5	220 (L) (1)	380/220 (A) 220 (L)	6·6 kV 380/220 (A)	±2 and±3
Ajman	50	230 (L) (1)	400/230 (A)	11 kV	(9)
Abu Dhabi	50	415/240 (A)	415/240 (A)	415/240 (A) (3)	(9)
UNITED KINGDOM (excluding Northern Ireland)	50±1	240 (L) (1)	415/240 (A)	22 kV 11 kV 6·6 kV 3·3 kV 415/240 (A)	±6
URUGUAY	50±1	220 (B) (L)	220 (B) (L)	15 kV 6 kV 220 (B)	±6
USA (4) Charlotte (North Carolina)	60±0·06	240/120 (K) 208/120 (A)	460/265 (A) 240/120 (K) 208/120 (A)	14·4 kV 7·2 kV 2·4 kV 575 (F) 460 (F) 240 (F) 460/265 (A) 240/120 (K) 208/120 (A)	+5, −2·5

City	Frequency				
Detroit (*Michigan*)	60±0·2	240/120 (K) 208/120 (A)	480 (F) 240/120 (H) 208/120 (A)	13·2 kV 4·8 kV 4·16 kV 480 (F) 240/120 (H) 208/120 (A)	+4, −6·6
Los Angeles (*California*)	60±0·2	240/120 (K)	4·8 kV 240/120 (G)	4·8 kV 240/120 (G)	±5
Miami (*Florida*)	60±0·3	240/120 (K) 208/120 (A)	240/120 (K) 240/120 (H) 208/120 (A)	13·2 kV 2·4 kV 480/277 (A) 240/120 (H)	±5
New York (*New York*)	60	240/120 (K) 208/120 (A)	240/120 (K) 208/120 (A) 240 (F)	12·47 kV 4·16 kV 480/277 (A) 480 (F)	(9)

Appendix 8

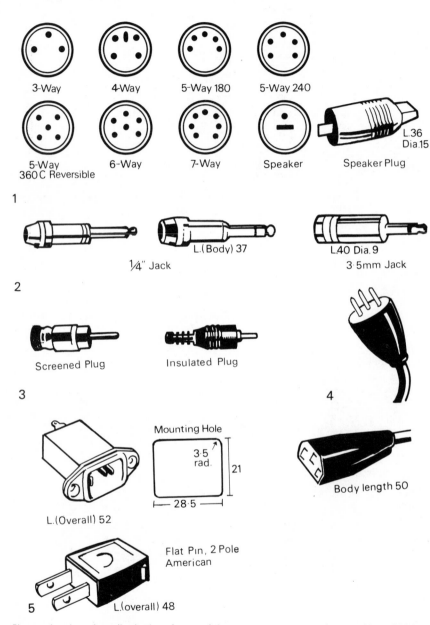

3-Way 4-Way 5-Way 180 5-Way 240

5-Way 360 C Reversible 6-Way 7-Way Speaker Speaker Plug

L.36 Dia.15

1

¼" Jack L.(Body) 37

L.40 Dia.9
3·5mm Jack

2

Screened Plug Insulated Plug

3 4

Mounting Hole

3·5 rad. 21

28·5

L.(Overall) 52

Body length 50

Flat Pin, 2 Pole American

5 L.(overall) 48

Plugs and sockets. A small selection of some of the more common connections used in multivision. 1, DIN connectors. 2, Jack connectors. 3, Phono connectors. 4, Cannon connector. 5, Socket; Plug (IEC standard); Mains connectors.

Glossary

Academy mask   Derived from the motion picture industry. Refers to the proportions of an exposed frame of a 35 mm movie film. Adopted because of the introduction of a soundtrack at the side of the film which spoiled the rectangular format. Academy mask is ·868 in wide by ·631 in high. Screen format is approx. 4:3.

Access time   In a storage device the time interval between demanding information and receiving it.

Acoustical power   The energy contained in a sound wave. Expressed as power per unit area (W per square m).

Alternate   In slide projector control, the transfer from one projection lamp to another without advancing the tray of the darkened projector.

Ambient light   The room-light level during projection. Should always be as low as possible.

Amplifier   A device for increasing power without altering the nature of that which is increased.

Analogue   A process in which numbers are represented by physical qualities such as, for example, voltage or current.

Animate   In slide projector control, a flashing sequence that gives an animation effect on the screen. A number of projectors within a single group will flash their lamps without advancing their slide trays. The animation pattern continues until a further cue stops it.

Animation   A rapid and sequential showing of a number of still images giving the appearance of movement to a single image.

Animation loop   The number of projectors used to give an animation effect on screen. When the last projector in the group has flashed its lamp, the cycle is repeated starting with the first projector.

Aperture, lens   The opening through which light is admitted to a lens. The width of the aperture relates directly to the transmission and focal length to determine the speed of a lens.

Aperture, slide   The internal measurement of the slide mount opening. Always slightly smaller than the camera plate aperture, and therefore smaller than the exposed film frame, in order to mask off any imperfections at the edge of the picture.

Aspect ratio   Width-to-height proportions of image area.

Audio frequencies   The frequencies which lie within the audible range. Approximately 16 Hz–16 000 Hz.

Audiovisual aids   Speaker support aids such as overhead projectors, flip charts, manually operated slide projectors. Not to be confused with 'Audiovisual' (synchronized slide/sound programmes). More properly termed Visual Aids.

Audiovisual format   A cassette tape track configuration designed to minimize crosstalk between control and audio channels. Tracks 1 and 2 audio; track 3 blank; track 4 control.

Autopresent unit   Automatic cartridge replay unit which includes electronics for the control of an audiovisual show. Will reset projectors at end of show.

Auxiliary   Output from a dissolve unit used to control equipment other than projectors. Relay contact closures provide switching facilities for lights, curtains, etc.

Azimuth error   An angular deviation between the axis of a record or playback head and a line at right angles to the direction of the travel of the tape.

Back projection   The projection of visual images onto a screen from the near side to that of the audience.

Baffle   In loudspeakers the isolation of sound coming from the front of the driver from the sound coming from the rear. This prevents the front and rear sounds from cancelling each other out and therefore reducing bass frequencies.

Batch   A particular quantity of filmstock which is manufactured at the same time. Each batch has an identification number since minor changes can occur.

Beaded screen   A front projection screen which because of a surface of bead-like formation reflects much of the light back along its axis.

Bit   A unit of information content which corresponds to a decision between one of two possible states. In storage devices: a unit of capacity. The capacity in bits of a storage device is equal to the logarithm to base two of the number of possible states of the device.

Blank slide   A slide through which light cannot pass, used at the beginning and end of shows to prevent light from reaching the screen. Special metal blanks can be purchased. An alternative is to mount foil or carbon paper in a normal glass mount.

Bloom Anti-reflective lens coating.

Brightness An attribute of visual perception which corresponds to more or less light from a particular source. Not to be confused with *luminance*.

Candela Unit of luminous intesity (S1 system).

Canned sequences Prerecorded and synchronized audiovisual sequences

Cartridge, continuous loop In tape replay, a recepticle which holds lubricated tape wound in a continuous loop, coating side outwards. Standards are mini-8 and NAB. Playing time of mini-8 is a maximum of 18 minutes at $3\frac{3}{4}$ ips. NAB cartridges are made in different sizes to accommodate 20, 30 or 40 minutes playing time.

Cels Cellulose acetate sheets. Transparent plastic sheets on which artwork is traced and painted for animation or slide registration photography.

Chop A fast picture change between two slide projectors in which the slide change occurs before the lamp is switched, allowing the shutter to chop off the light.

Chromium dioxide (C_rO_2) A magnetic particle coating for sound (and video) recording tape. Used in cassettes. Chromium dioxide is made in a high temperature/pressure reaction. Characteristics are in its slender, needle-shaped and uniform particles in which the magnetic axis is nearly the same as their geometric axis. C_rO_2 tape requires higher bias currents in the record head because of its high coercivity.

Clock frequency The master frequency of timing pulses which govern schedule of operation.

Clock pulse A pulse applied to logical elements to effect logical operations.

Clock track A recording on tape of encoded signals which when decoded display the 'real-time' of that recording. Used for aiding accurate synchronization of sound and picture in multivision, but is not used directly for projector control. Any selected 'tick' of a clock track when decoded will give a read-out of the time in minutes, seconds and half seconds (or $\frac{1}{10}$ second).

CMOS Complementary Metal Oxide Substrate.

Coercivity (H_c) The resistance of a recording tape to magnetizing or demagnetizing. A measure of the force needed to remove the remanence of the magnetic coating.

Colours, complementary Those colours which are obtained by subtracting the three primary colours from the visible spectrum. Thus magenta (minus green); cyan (minus red); yellow (minus blue).

Colour temperature Measured in kelvins, the colour of a light source when related to a theoretically perfect (black body) source of radiant energy. Daylight film is balanced for 5600 K. Tungsten light film is balanced for 3200 K. The higher the number of kelvins the more bluish the light colour. The lower the number of kelvins the redder the light colour.

Condensers Lens elements used for spreading and directing projection light into the slide aperture.

Contrast In a photographic image the difference between the densities of the most exposed and least exposed areas.

Cos⁴ θ law The illumination E in the focal plane of an image point at an angle θ to the optical axis is proportional to $\cos^4 \theta$. Unless modified, wide-angle lenses give reduced image illumination at the edge of the projected image according to the $\cos^4 \theta$ law.

Crosstalk The ratio, expressed in decibels, between two voltages occurring at the output terminals of two adjacent tape tracks. A measurement of the unwanted signal on one track which has been produced by the recorded signal on the other track.

CSR Capability/size ratio. Applied by manufacturers to equipment to emphasize portability.

Cue chain A sequence of cues that are executed by a single pulse.

Cue sheet An instruction sheet used by a person synchronizing or programming an audiovisual show. Contains information on cue times, cue numbers, dissolve rates and special effects.

Cue tone A burst of sinusoidal tone used for cueing a slide projector to display a slide at a given point in time. Also known as sync pulse.

Cut In slide control, a fast dissolve effect achieved by switching the lamps of the projectors. The lamp is allowed to be fully extinguished before the slide change takes place.

Cycle time Of a magnetic memory, the minimum time interval between the starts of successive read-write cycles.

Decoder (1) In computers, a network in which several inputs are excited together to produce a single output; (2) In audiovisual, specifically, a device that interprets pulsed signals and outputs commands to dissolve units or projectors.

Decrement Stepping through cues in a reverse direction with a programmer.

Demodulator A device for separating the original information from a modulated signal wave. In multivision control a demodulator may also carry out other duties besides this primary function.

Diallable colour head Adjustable colour filtration system with which you can control the degree of magenta, cyan or yellow filtration for duplication or printing of photographic material.

Diaphragm The moving portion of a microphone which is activated by sound waves.

Diapositive An alternative name for a transparency or slide.

Diascope Early name for a slide projector.

DIN Deutscher Industrie Normenauschuss (German Standards Authority).

Diode An electronic analogue device allowing current to flow in one direction only. Silicon diodes give a forward voltage drop of 0·7 volts.

Dissolve loop The number of projectors in a dissolving sequence. For instance, with three projectors dissolving from A to B to C and back to A.

Slide advance occurs on the darkeded projector after each dissolve.

Dissolve unit   An apparatus for simultaneously dimming down the light on one slide projector and fading up the light on another slide projector.

Do-next loop   A method of programming certain memory encoders to enable sequences to be repeated.

Dope sheet   A sheet of instructions used by a rostrum cameraman.

Drum   A rotary slide tray.

DTL   Diode Transistor Logic.

Dubbing   Duplicating or re-recording of a soundtrack for purposes of editing, synchronizing or making copies.

Dump   To store the temporary memory contents of a programmer on magnetic tape.

Dynamic microphone   A moving coil microphone.

Eggboxing   In rear projection, the system used for dividing individual screen areas to avoid overspill of projected images from one screen to another.

Elcaset   Large format cassette tape. Uses tape which is twice width of standard Philips cassette. Tape width is 6·35mm. Tape speed 9·5cm/s ($3\frac{3}{4}$ips).

Electrostatic copy holder   A method of holding artwork in a perfectly flat position for photography.

Electrostatic microphone   A capacitor microphone.

Equalization   In audio recording, electronic circuitry for achieving an overall flat frequency response throughout the audible frequency range.

Engineering pulse   A single control signal at the beginning of an audiovisual programme tape for removing blank slides from projectors. Not needed on systems which do this automatically when a start signal is given.

Epidiascope   A projector which combines the functions of both *episcope* and *diascope*, projecting either opaque or transparent originals.

Epi-illumination   Illumination of an opaque original for projection.

Episcope   A visual aid which projects opaque originals.

Eye-letting   In the manufacture of screens small holes with protective metal rings placed in the webbed border of the screen. Used for 'flying' the screen with strong cord.

Fade rate   The time taken for a projector lamp to fade up or down. Alternatively, the number used to denote a particular rate of fade.

Field size   The artwork area which is actually to be photographed.

Field size indicator   On a copy stand a strip of write-on material mounted on the column on which the field size is indicated by a pointer mounted on the camera. More elaborate systems show the field size as a read-out on a digital display.

Fill   To load a program into a computer memory from an external source, e.g. magnetic tape.

Follow focus cam   A cam which is cut to correspond to the focusing curve of an individual lens. Used to automatically focus a rostrum camera as it is raised and lowered.

Foot-candle   Unit of illumination. It is the illumination on one square foot of surface on which there is an evenly distributed luminous flux of one lumen.

Foot-lambert   'Apparent foot-candle'. A unit of brightness. Equal to 1/ft candle per square centimetre. Equal also to the uniform brightness of a perfectly diffusing surface emitting (or reflecting) light at the rate of one lumen per square foot. The average brightness of, say, a screen surface in foot lamberts is equal to the illumination in foot candles *times* the reflecting factor of the surface.

FX   Sound effects.

Gain   An increase (or decrease) in electrical signal strength. Alternatively, a measure of transmitted brightness given by screen material.

Gate   (1) The projector mechanism that holds a slide in position during projection; (2) An electronic circuit with more than one input but only one output.

Glow   In a projector lamp, the low level of current which keeps the filament in a red-hot state, thereby preventing high current surges when the lamp is switched on from cold. Used in some systems to lengthen lamp life.

Generation   Each successive stage in the duplication of reproductive material.

Graticule   *See* Reticle.

Hardware   The physical parts of electronic equipment.

Helical focusing   Movement of a lens for focusing in which the lens rotates while moving axially. As opposed to 'rectilinear focusing' in which the lens does not rotate. Helical focusing is used with nearly all slide projector lenses.

Home   The zero (start) position of slide projector trays.

Homing   The resetting of projectors to the zero position.

Homing time   The time taken for a complete revolution of a slide tray. On a Kodak S-Av 2000 should not be greater than 2 minutes 7 seconds.

Housekeeping functions   In multivision control the secondary functions of control equipment such as the display of lamp fail and slide number information or the issuing of pulses to give the dimming up and down of house-lights.

Hot spot   In rear projection the brighter centre part of the projected image. Caused by a combination of a 'high gain' screen material and short focal length lenses.

Illuminance   A measurement of the amount of light falling on a given subject. Measured in lux (S1 system).

Impedance   The opposition a circuit offers to the flow of electrical current.

244

Increment Stepping through cues in a forward direction with a programmer.
Inhibit step A command given to an automatic slide projector to leave the slide in the gate after the lamp has been faded down.
Invertor An electronic component that gives an output that is the inverse of the input.

Keystone distortion The distortion of a projected image caused by tilting projectors off the ideal 90° to the screen plane axis. Some keystone distortion is inevitable when using more than one projector to cover a single screen area. In practice this is masked off at the edges of the screen.
Keystone eliminator A metal bracket on top of portable tripod screen for tilting screen surface towards audience. Useful only for overhead projectors or single slide projectors.

LED Light Emitting Diode.
Level The volume of an electrical signal. In sound recording 'correct' level will lie between noise level and distortion level.
Line-up Initial alignment of projectors when setting-up for showing an audiovisual presentation.
Line-up slides may be used to achieve accurate overlapping of two (or more) projectors. A line-up switch on replay equipment enables both projector lamps to shine at the same time.
Line-up slides Transparencies mounted in registration mounts for purposes of accurately aligning projectors. Usually showing a grid pattern for measuring definition and possibly tonal or colour saturation scales. Attractive line-up slides can be produced in pairs with each slide showing, say, 50% of a company logo. The two images projected in register show the complete symbol.
Lith mask A mask produced by photographing artwork on high contrast film for mounting with normal transparency in slide mount. Used for obtaining unusual slide formats, e.g. a head shape or a company logo. In optical printing it is used for producing composite images on a single slide.
Lumen Unit of luminous flux.
Luminance A photometric quantity for measuring light intensity. It is independent of subjective sensation. The luminous intensity of a surface (measured from a given direction) per unit of projected area viewed from the same direction.
Luminous flux Total visible light energy emitted by a source in unit time.
Lux The practical unit of illuminance in the S1 unit system of metric measurement. 1 lux is 1 lumen per square metre.

Memory encoding The programming of an audiovisual display by inserting commands into a temporary computer memory. Does not synchronize an audiovisual show unless a time code is linked to each cue.

Microprocessor   A miniature computer system containing a memory store and a processing unit.

Modulation   The variation of the characteristic of a wave in sympathy with the characteristic of another wave. For instance, the superimposition of an audio signal on a carrier wave of higher frequency.

Multiplex   The simultaneous transmission of several functions, for example, frequency, amplitude or waveshape, without loss of identity of any function.

Multiplexing   The operation of a single path for the transmission of more than one set of signals simultaneously. *See* 'Time division multiplex'.

Multiplex speed   The speed at which a multiplex signal is transmitted. Poor recording media such as 8 mm magnetic stripe require multiplex signals to be 'sent' at half speed. High quality media can accept multiplex at double speed.

Newton's rings   In glass-mounted transparencies the ring-like effects caused by the celluloid side pressing unevenly against the glass.

Optical axis   Theoretical line joining centres of curvature in lens elements in an optical system.

Pantograph table   A small table or area attached to the side of an animation or registration stand. A pointer attached to the main animation table indicates its position in relation to previous positions.

Pegs (1)   On the base of a slide magazine the plastic divisions between slide positions which are engaged by the tray transport lever to step the projector.

Pegs (2)   The registration pins on a peg bar.

Peg bar   A metal strip with three pegs for holding artwork on an animation table in exact position.

Phon   Unit of subjective loudness. At 1000 Hz is equal to decibels at which frequency only phon scale is additive. At other frequencies phons must be changed to *sones* to find units of comparative loudness.

Platen   A hinged sheet of glass for pressing artwork into a flat plane for photographing.

Point size   An index used in printing. Refers to the height of the standard backing block on which the raised letter appears. In European point system there are 72 points to the inch.

Presentation unit   An audiovisual show presenter containing a dissolve unit, decoder, cassette deck, audio amplifier and controls.

Pressure gradient   The difference between two successive points in a sound wave. Principle used in directional microphones.

Pulse   A variation of a quantity whose value is normally constant.

Puncher   A machine for putting registration holes into artwork and cels.

Q Abbreviation of 'cue'.

Q-slide Term used for the continuous tone method of encoding. A slider control is used to programme the show.

Ramp Electronic control signals which are fed to a triac for increasing or decreasing lamp intensity.

Real time The actual viewing time of an audiovisual programme.

Random access projection A specialized visual aid giving the facility of projecting a slide from any position in one or more projectors in a very short access time.

Registration pins Fixed-pin registration is found in high quality optical printers. Part of the camera mechanism for holding the film in exact register during exposure. Also in registration slide mounts for positioning the transparency in exact register to others on similar mounts.

Remanence (Br) A characteristic of magnetic tape. The amount of magnetization remaining after a magnetic field has been applied.

Reticle On a rostrum-mounted camera, the ground glass viewing screen containing a grid which shows slide formats and field sizes. Also sometimes referred to as the 'graticule'.

Root mean square value RMS. The square root of the average value of the squares of the instantaneous values during a complete cycle. Applies to amperes, volts, watts or other recurring variable quantities.

RTL Resistor Transistor Logic.

Saturation level (Bs) A characteristic of magnetic tape. The point at which no further magnetization is possible.

Screen tint A tint given to rear-projection screen material to cut down the unwanted effects of ambient light.

Servo system An automatic control system in which the output element or quantity closely follows the input to the system. It will have at least one feedback loop which corrects successive inputs having measured the difference between actual and desired output. An example in audiovisual control is in the dimming of xenon projectors by normal dissolve units.

Shadow board A device mounted beneath a rostrum camera lens which contains an appropriate aperture but which prevents reflections travelling between camera and artwork.

Sine wave (Sinusoidal wave) A wave whose displacement is the sine (or cosine) of an angle proportional to time or distance or both.

Silica gel Chemically activated silicon dioxide which is used to absorb moisture in the storage of slides or electronic equipment.

Slide lift height The amount by which a slide is lifted by the slide lift arm from the projector gate. On S-AV 2000 should be 3·5mm. If incorrect, jamming of slide tray can occur.

Snap A snap change is a shutter-operated image change from one projector to another. Achieved by fitting a solenoid-controlled shutter to a projector.

Software Borrowed computer term which in audiovisual usage means the slide/sound programme. Can be confusing now that computers (which are themselves programmed with software) are used for audiovisual control.

Solenoid A coil for producing a magnetic field. Sometimes includes a rod which can move along the axis of the coil under the influence of its magnetic field. Used for remote control of electromechanical devices.

Speaker support sequences Slide sequences without sound accompaniment for display while a presenter is speaking.

Square wave A wave that has a square profile when examined by an oscilloscope. It is a periodic wave which alternately assumes, for equal lengths of time, one of two fixed values.

Stacking stand Also known as 'twinning stands'. A piece of equipment used to stack projectors one above the other. These may be either simple tripod-type stands or the more professional kind which will adjust the direction of the projector beam. The second type are available for stacking either two or three projectors one above the other.

Stepping Advancing a slide projector one slide at a time. Hence 'forward stepping' and 'reverse stepping'.

Storyboard The script of a programme with accompanying sketches of the proposed visuals.

Superimposition In projection, the overlay of one image on to another by projecting a second slide on the same screen area. With many dissolve-type systems the superimposing slide must be taken off-screen before the show continues.

Synchronization The matching up of sound and visuals. The synchronization of slides with a prerecorded soundtrack is used in all audiovisual shows.

Task light The light needed by an audience to take notes.

TTL (1) Transistor Transistor Logic (electronics); (2) Through The Lens (photography).

Thermal cut-out A device for switching off power to a projector because of overheating.

Throw The distance travelled by a projector beam from lens to screen.

Thyristor dimmer Circuitry which passes on to a lamp a defined fraction of the AC power available from the supply. Used in dissolve units.

Time division multiplex The transmission of two or more signals over a common path using different time intervals to carry information.

Transmission gate An electronic (CMOS) bidirectional switch controlled by voltage. When voltage is high it is enabled, when low it is disabled.

Tray transport lever On a slide projector the arm which stops the slide tray.

Ultrasonic cleaning A method of cleaning mechanical and electrical apparatus (especially useful for projectors) by dipping them in a chemical bath through which high frequency sound waves are passed. Every particle of grease and dirt is removed. Mechanical parts should afterwards be lubricated.

Wash out   Loss of screen image caused by excessive ambient light.
Write-on slide   A temporary slide used during production on which you can indicate the final photographed image.

Zoom lens   A lens with a variable focal length. As a projection lens will provide varying sizes of picture at a given distance.

Index

250

251